WEIGHING DESIGNS

FOR

Chemistry
Medicine
Economics
Operations Research
Statistics

STATISTICS

Textbooks and Monographs

A SERIES EDITED BY

D . B . O W E N , Coordinating Editor
Department of Statistics
Southern Methodist University
Dallas, Texas

PAUL D. MINTON
Virginia Commonwealth University
Richmond, Virginia

JOHN W. PRATT
Harvard University
Boston, Massachusetts

OTHER VOLUMES IN PREPARATION

WEIGHING DESIGNS

FOR

Chemistry • Medicine • Economics
Operations Research • Statistics

KALI S. BANERJEE

Department of Statistics & Computer Science
University of Delaware
Newark, Delaware

MARCEL DEKKER, INC. New York

MARCEL DEKKER, INC.
270 Madison Avenue, New York, New York 10016

LIBRARY OF CONGRESS CATALOG CARD NUMBER: 75-194

ISBN: 0-8247-6287-8

Current Printing (last digit):
10 9 8 7 6 5 4 3 2 1

PRINTED IN THE UNITED STATES OF AMERICA

To My Wife Niti

CONTENTS

The weighing problem originated in a casual illustration
furnished by Yates [Journal of the Royal Statistical Society,
Supplement 2, pp. 181-247 (1935)]. This illustration later led
to a precise formulation of the weighing problem by Hotelling
[Annals of Mathematical Statistics, vol. 15, pp. 297-306 (1944)].
Over the years the problem has attained a distinctive growth,
has branched out in different directions, and has acquired
meanwhile the status of a problem in the design of experiments.
The problem has also become associated with the name of Hadamard
and has given noticeable impetus to research in the extension of
the Hadamard determinant problem.

Besides being helpful in routine weighing operations to determine the weights of light objects, the results of research in
this area appear to have the potential of being useful in chemical, physical, biological, economic, and other sciences. Weighing designs are, in fact, applicable to any problem of measurements in which the measure of a combination is expressible as a
linear combination of the separate measures with numerically
equal coefficients [Mood, Annals of Mathematical Statistics,

vol. 17, pp. 432-446 (1946)]. It is thus felt that there is
perhaps a necessity for introducing the problem in general
terms to research workers in all applied sciences, so that they
might have an idea of what the problem is about and what results
are available. It is this feeling that has led to the preparation of this monograph.

This volume contains four chapters. Chapter 1 starts from
the original example of weighing operations (the example that
gave rise to what has become known as the weighing problem) and
provides, in the sequel, the basic structure of the weighing
problem with reference to the statistical model characterizing
the problem.

Chapter 2 is on the chemical balance (two-pan balance)
problem, and Chapter 3 is a discussion of the spring balance
(one-pan balance) problem. These two chapters furnish charac-
terizations of efficient chemical balance and spring balance
weighing designs, respectively.

Chapter 4 deals with miscellaneous issues connected with
the weighing problem, and contains a brief descriptive review
of the achievements in this area. This chapter is the largest,
although it merely touches on the various aspects of weighing
designs dealt with to date.

Keeping in view the fact that this monograph may be pe-
rused by research workers in sciences other than statistics,
three appendices have been added to meet the needs of such
research personnel.

Appendix A is on the general linear hypothesis model and
the method of least squares, which provide a background for
estimation in weighing designs. This appendix gives only those
of the few basic results on least squares that are needed in
connection with the weighing problem. Supplementary references,
however, have been cited at the end of this appendix, in case
the reader would like to consult a detailed study of the subject matter.

In Appendix B weighing designs are illustrated with numerical examples along with a description of the steps we need to take in the use of weighing designs. The required arithmetic is set out in detail. This appendix contains, in addition, two working rules meant to provide the estimates of weights as a routine measure.

Appendix C contains a short list of efficient chemical balance and spring balance weighing designs. The variance factors for the estimated weights are furnished at the bottom of each of these designs in order to give an idea of the efficiency of the design. Although not exhaustive, this list serves the purpose of providing ready reference to weighing designs involving a smaller number of objects.

A few exercises have been provided. These exercises, many of which are relatively simple, cover the basic background material.

Details of mathematical derivation have been avoided in this monograph as far as possible. Instead, a complete list of references has been cited at the end. This list contains, as far as is known to this author, all papers that have a bearing on the weighing problem. If any paper has not been mentioned, it is because this author is not aware of the contribution. The author would feel grateful if such ommisions were brought to his attention.

ACKNOWLEDGMENTS

Of numerous well-wishers in India and in the U. S. A. to
whom I owe a debt of gratitude, I would like to mention the
names of Professor R. C. Bose, now at Colorado State University,
Professor J. C. Kiefer and Professor W. T. Federer of Cornell
University, Professor J. Wolfowitz of the University of Illinois,
Professor H. C. Fryer of Kansas State University, Professor D.
B. Owen of Southern Methodist University, Professor D. E. Lamb
and Professor J. F. Leathrum of the University of Delaware, and
Dr. R. N. Carr of Schering Corporation.

I would also like to mention the name of my youngest sister,
Bakul, who helped me with the weighing.

WEIGHING DESIGNS

FOR

Chemistry
Medicine
Economics
Operations Research
Statistics

Chapter 1

FORMULATION OF THE WEIGHING PROBLEM

1.1 INTRODUCTION

In this volume, we present a descriptive summary of the
basic results obtained in the context of the weighing problem
and weighing designs. We start from the origin of the problem
and make note of subsequent developments as we proceed.

1.2 ORIGIN OF THE PROBLEM

1.2.1 Original Example of Yates

In furnishing an illustration of "independent factors" in
complex experiments, that is, factors that do not interact, Yates
[78] considered the following problem: A chemist has seven light
objects to weigh, and the scale requires a zero correction. The
obvious technique would be to weigh each of the seven objects
separately and to make an eighth weighing with no object on the
scale so that the zero correction could be determined. Thus,

1

to determine the weight of each object, one would take the difference between the readings of the scale when carrying that object and when empty. Assuming that systematic errors are non-
existent and that the errors are random, we denote the standard
error of each weighing by σ, and the variance by σ^2. Given
these assumptions, the variance of the estimated weight is $2\sigma^2$,
and its standard error is $\sigma\sqrt{2}$.

As an improvement over the customary technique, Yates [78]
suggested that the objects be weighed in eight combinations according to the following scheme:

Weighing no.	Objects weighed
1	$a + b + c + d + e + f + g = y_1$
2	$a + b \quad\;\; + d \qquad\qquad\qquad = y_2$
3	$a \quad\;\; + c \quad\;\; + e \qquad\qquad = y_3$
4	$a \qquad\qquad\qquad\quad + f + g = y_4$
5	$b + c \qquad\quad + f \qquad = y_5$
6	$b \qquad\quad + e \quad\;\; + g = y_6$
7	$c + d \qquad\quad + g = y_7$
8	$d + e + f \qquad = y_8$

$$(1.1)$$

In this scheme, each object is weighed four times in the
different combinations. In the four weighings of a given object,
every other object is included twice. The remaining four weighings, that is, weighings without the object, also include every
other object twice. If the readings from the scale are denoted
by y_1, y_2, $\ldots$, y_8, the weight of any object a is given by

$$a = \frac{y_1 + y_2 + y_3 + y_4 - y_5 - y_6 - y_7 - y_8}{4}$$

A similar expression is obtained for each of the other objects. It is evident from the above expression that the weight
of any particular object is found by adding together the four

equations containing it, subtracting the other four, and dividing the algebraic sum by 4. It can further be seen that the constant bias, if any, which affects each observation y, cancels
out in the algebraic sum. This bias may be due to the balance
requiring a zero correction, or it may be the result of several
"environmental characteristics."

Since the variance of a sum of independent observations is
the sum of the variances, the variance of a by this improved
technique is $\sigma^2/2$, which is 1/4 that of the customary method.

1.2.2   Improvement Suggested by Hotelling

Hotelling [49] suggested that a further improvement would
be possible if Yates' procedure were modified by placing in the
other pan of the scale those objects not included in the weighing. Calling the readings Z_1, Z_2, ..., Z_8, we can write the
scheme of weighing operations [interchanging c and d of
scheme (1.1)] as

$$
\begin{aligned}
a + b + c + d + e + f + g &= Z_1 \\
a + b + c - d - e - f - g &= Z_2 \\
a - b - c + d + e - f - g &= Z_3 \\
a - b - c - d - e + f + g &= Z_4 \\
-a + b - c + d - e + f - g &= Z_5 \\
-a + b - c - d + e - f + g &= Z_6 \\
-a - b + c + d - e - f + g &= Z_7 \\
-a - b + c - d + e + f - g &= Z_8
\end{aligned}
\qquad (1.2)
$$

From these equations,

$$
a = \frac{Z_1 + Z_2 + Z_3 + Z_4 - Z_5 - Z_6 - Z_7 - Z_8}{8}
$$

A similar expression is obtained for each of the other unknowns. The variance of each unknown by this method is $\sigma^2/8$.
The standard error is half that by Yates' method, or a quarter

of its value by the direct method of weighing each object separately. Here also, the bias cancels out.

It should be pointed out that, in the example furnished by
Yates, the objects were placed in only one pan of the scale.
One pan is used when the balance is of the spring balance type.
But in the improvement suggested by Hotelling, both pans of the
scale were used. This, however, is possible only in a chemical
balance (see Sec. 4.4).

The design principle involved in the above method can perhaps be better illustrated with reference to a simpler example.
Let us suppose that it is necessary to find the unknown weights
of two light-objects a and b and that the scale to be used
is corrected for bias. If the two objects are weighed together
in one pan of the scale, and also in opposite pans, the equations for the unknown weights are

$$a + b = y_1 \quad a - b = y_2$$

where y_1 and y_2 denote the readings from the scale. From
these equations, we get

$$a = (y_1 + y_2)/2 \quad b = (y_1 - y_2)/2$$

If σ^2 is the variance of an individual weighing, the variance
of a and b, by this method, is $\sigma^2/2$. The error for both
estimates, therefore, is $\sigma/\sqrt{2}$. Thus, with only two weighing
operations, it is possible to obtain the same standard error,
$\sigma/\sqrt{2}$, for both objects. If, however, the objects were weighed
separately, twice each, four weighing operations would be necessary to obtain this standard error for both. Weighing the objects in combination therefore reduces the number of weighing
operations by half.

This example amply illustrates the fact that "when several
quantities are to be ascertained there is frequently an oppor-
tunity to increase the accuracy and reduce the cost by suitably

combining in one experiment what might ordinarily be considered
separate operations" (Hotelling [49]).

1.3   STATISTICAL MODEL CHARACTERIZING THE PROBLEM

In the light of the above illustrations, Hotelling [49]
gave a precise formulation of the problem of weighing designs,
which can be summarized as follows:

The results of N weighing operations to determine the individual weights of p light objects fit into the general lin-
ear hypothesis model $\underline{Y} = X\underline{\beta} + \underline{\epsilon}$, where $\underline{Y}$ is an N × 1 random
observed vector of the recorded weights; $X = (x_{ij})$ (i = 1, 2,
..., N; j = 1, 2, ..., p) is an N × p matrix of known quantities,
with x_{ij} = +1, -1, or 0, if in the ith weighing operation the
jth object is placed, respectively, in the left pan, in the
right pan, or in none; $\underline{\beta}$ is a p × 1 vector (p ≤ N) representing the weights of the objects; $\underline{\epsilon}$ is an N × 1 unobserved random
vector such that $E(\underline{\epsilon}) = \underline{0}$ and $E(\underline{\epsilon}\underline{\epsilon}') = \sigma^2 I_N$.

Consistent with the signs that the elements x_{ij} can take,
the record of the ith weighing is taken as positive or negative,
depending on whether the balancing weight is placed in the right
pan or the left.

We shall call X the "design matrix." When X is of full
rank, that is, when [X'X] is nonsingular, the least-squares
estimates of the weights are given by $\hat{\underline{\beta}} = [X'X]^{-1}X'\underline{Y}$, where X'
is the transpose of X. The covariance matrix of the estimated
weights is given by $cov(\hat{\underline{\beta}}) = \sigma^2[X'X]^{-1} = \sigma^2 C$; c_{ii}, which is
the ith diagonal element of C, represents the variance factor
for the ith object.

We have mentioned above only a few of the results that are
obtainable under the Gauss-Markoff linear hypothesis model.

These results and all others in the theory apply, in general,
to the model for weighing designs.

In particular, when X is a square matrix of full rank,
the least-squares estimates are given by $\hat{\underline{\beta}} = [X'X]^{-1}X'\underline{Y} = X^{-1}\underline{Y}$.
On the other hand, if we solve for $\underline{\beta}$ in $\underline{Y} = X\underline{\beta}$, $\underline{\beta}$ is also
given by $X^{-1}\underline{Y}$. This means that the least-squares estimates
are the same linear functions of the observed y's as the true
parameters would be of the true y's. If X is an orthogonal
matrix in the sense that $[X'X] = NI_N$, then $\hat{\underline{\beta}} = X^{-1}\underline{Y} = [X'\underline{Y}]/N$
(see Appendix A).

1.4   TWO TYPES OF PROBLEMS

Two distinct types of problems arise in practice, one with
reference to the spring balance and the other to the chemical
balance. In the spring balance problem, the elements x_{ij} are
restricted to values of +1 or 0, while in the chemical balance
problem these elements are either +1, -1, or 0.

It should be mentioned that the designs used in the weighing problem are applicable to a variety of problems. In fact,
they are applicable to any problem of measurements, in which the
measure of a combination is a known linear function of the separate measures with numerically equal coefficients. In the interest of simplicity, the problem has been discussed in the language of weighing operations [56], and we shall also adopt this
language.

1.5   HOTELLING'S LEMMA

Denoting by A the determinant of $[X'X] = (a_{ij})$ (i,j =
1, 2, .. , p), Hotelling [49] proved the following lemma:

<u>Lemma</u>. If a_{12}, a_{13}, ..., a_{1p} $(= a_{21}, a_{31}, ..., a_{p1}$, respectively) are free to vary while the other elements of A remain fixed, the maximum value of A/A_{11} is a_{11} and is attained when and only when $a_{12} = a_{13} = \cdots = a_{1p} = 0$, where A_{11} is the minor of A obtained by deleting the first row and column.

From this lemma, it is evident that the variance of $\hat{\beta}_1$, namely $\sigma^2(A_{11}/A)$, cannot be less that σ^2/a_{11} and the variance will reach this value only if the experiment is arranged such that the elements after the first row and column of A are all zero. This minimum value, σ^2/a_{11}, is attained when the first column of X is orthogonal to all the others. It is also clear that the minimum minimorum [49] of the variance will be reached not only if the first column of X is orthogonal to all the others, but also if it consists entirely of +1 and -1 elements, so that $a_{11} = N$. The maximum possible value that a_{11} can take is N. The value of this minimum minimorum is thus equal to σ^2/N.

It is evident from the lemma and the above discussion that this minimum minimorum of the variance will be reached with respect to all the estimates $\hat{\beta}_i$ $(i = 1, 2, ..., p)$ if the design matrix X is orthogonal in the sense that $[X'X]$ is diagonal, with N on the diagonal.

1.6   ILLUSTRATIONS

1.6.1   Chemical Balance

The design matrix X in the scheme of weighings in (1.2) is an 8×7 matrix whose elements are +1 and -1. The seven columns are orthogonal. If a column of +1's is added, we have an 8×8 orthogonal matrix as shown below (1 following the plus or minus sign is not indicated):

$$X = \begin{bmatrix} + & + & + & + & + & + & + & + \\ + & + & + & + & - & - & - & - \\ + & + & - & - & + & + & - & - \\ + & + & - & - & - & - & + & + \\ + & - & + & - & + & - & + & - \\ + & - & + & - & - & + & - & + \\ + & - & - & + & + & - & - & + \\ + & - & - & + & - & + & + & - \end{bmatrix} \qquad (1.3)$$

Scheme (1.2) was designed to find the weights of seven light objects on a chemical balance, which was known to have a bias. If the bias is taken as an additional object whose weight is to be determined, and if the first column is made to correspond to the bias, design matrix (1.3) is suitable for estimating the weights of the seven objects and the bias and is, in fact, the best design for the purpose by virtue of Hotelling's lemma.

Thus, scheme (1.3) can be used as a design for finding the weights of eight different objects, if the balance is free of bias, or the weights of seven objects, if the balance has a bias.

The rows of matrix (1.3) indicate how to combine the objects in the weighing operations, and the columns refer to the objects to be weighed. Matrix $[X'X]$ is diagonal, having eight on its diagonal. The diagonal form of the matrix reduces the solution of the normal equations (pertaining to the least-squares estimates) to the trivial task of dividing by 8 each of the numericals occurring on the right-hand side of the normal equations. That is the estimates are given by $\hat{\beta} = [X'\underline{Y}]/8$ (see Appendix B).

In general, if the results of the weighings are denoted by y_1, y_2, $\ldots$, y_8, the least-squares estimates of the unknowns are given by $(\ell_1 Y_1 + \ell_2 Y_2 + \cdots + \ell_8 Y_8)/8$ where the ℓ's are +1 or -1 in accordance with a row of signs of the transpose of matrix (1.3) or in accordance with a column of signs of matrix (1.3).

Thus, the estimate of a, for instance, is the same as given by
Hotelling [49].

In the above illustration, the inverse of matrix [X'X] is
also of the diagonal form, the elements in the diagonal being
1/8. The variance of each of the estimates is therefore $\sigma^2/8$.
With eight weighing operations, it is possible to obtain the
least possible variance of $\sigma^2/8$ for each of the eight objects.
If each object were weighed separately, 64 weighing operations
would be needed to arrive at this precision.

From the above, it is evident that such a saving of weighing operations is possible not only in the case of eight objects,
but also with any number N of objects, provided an orthogonal
matrix of order N of the above type, with $\pm$ 1 as its elements,
exists. Such matrices serve as the chemical balance designs of
maximum possible efficiency, since the minimum minimorum of the
variance is attained by each object.

1.6.2   Estimation of Error Variance

When scheme (1.3) is used for determining the weights of
eight objects from eight weighing operations, no degrees of freedom are left for estimating the error variance. If we have
eight objects whose weights are to be determined, and if the
error variance must also be estimated from the experiment, scheme
(1.3) can be repeated twice so that the resultant design matrix
is of dimensions 16 × 8. In this case, 16 - 8 = 8 degrees of
freedom will be left for estimating the error variance.

However, the same scheme may be used for determining the
weights of less than eight objects. If, for instance, there
are five objects whose weights are to be determined, any five
columns of scheme (1.3) can be used as the weighing design. In
that case, [X'X] will be of dimensions 5 × 5 and will be diagonal, with eight on its diagonal. The variance of each of the
five estimated weights will be $\sigma^2/8$, and 8 - 5 = 3 degrees of

freedom will be left for estimating the error variance (see Appendix B).

1.6.3   Estimation of Bias

If the balance has a bias, the first column of scheme (1.3), which consists entirely of +1's, can be made to refer to the bias. The seven other columns refer to the seven objects to be weighed. Estimates of the weights of the seven objects and the bias are obtained as above. The estimated bias will have the same precision as that of the estimated weight of each of the seven objects. However, if in a scheme like that of (1.3) the signs of the first column are not all plus, the same scheme can be prepared to suit the estimation of bias. The signs of the first column can be made positive with corresponding changes of signs in the respective rows. The first column thus prepared can be made to refer to the bias. It can easily be seen that changing the signs in this way does not alter the value of the variance factors.

1.6.4   Spring Balance

From the above, it is evident that orthogonality between columns of the design matrix X requires that the signs of the elements of X be both positive and negative, which is possible only if both pans of the balance are used. Thus, design matrices with the minimum minimorum of the variance for each object are not possible for spring balance measurements or any other measurements in which it is not possible to ensure that the quantities read-off are differences.

However, for $N = p = 3$, Hotelling [49] found the following design matrix to be the most efficient (he presumably found this by trial):

$$X = \begin{bmatrix} 1 & 1 & 0 \\ 1 & 0 & 1 \\ 0 & 1 & 1 \end{bmatrix} \qquad\qquad\qquad (1.4)$$

Here, $[X'X]$ and $[X'X]^{-1}$ are given, respectively, by

$$[X'X] = \begin{bmatrix} 2 & 1 & 1 \\ 1 & 2 & 1 \\ 1 & 1 & 2 \end{bmatrix} \qquad [X'X]^{-1} = \begin{bmatrix} 3/4 & -1/4 & -1/4 \\ -1/4 & 3/4 & -1/4 \\ -1/4 & -1/4 & 3/4 \end{bmatrix}$$

Estimates $\hat{\beta}_1$, $\hat{\beta}_2$, and $\hat{\beta}_3$ are given, respectively, by $(y_1 + y_2 - y_3)/2$, $(y_1 - y_2 + y_3)/2$, and $(-y_1 + y_2 + y_3)/2$. The variance of each unknown is $3\sigma^2/4$.

This design, as we shall see later, is a special case of a series of designs, L_N, of Mood [56], which in turn has been recognized [1] to be a special case of balanced incomplete block designs (BIBD). Solution of the normal equations, in the general case of a BIBD, is equally simple and can be reduced to a simple routine [9].

1.7   HOTELLING'S CALL FOR FURTHER MATHEMATICAL RESEARCH

Reaching the point described above, Hotelling [49] furnished several additional illustrations explaining the spring balance problem and observed that there was a need for further mathematical research in this direction. In response to his call, there have since been several articles on the subject (Kishen [53], Mood [56], Rao [64], Plackett and Burman [59], Kempthorne [50], Banerjee [1-19], Raghavarao [60-63], Zacks [83], Beckman [21, 22], Sihota and Banerjee [67], Hazra and Banerjee [47], Hazra [46], Dey [31-33], Kulshrestha and Dey [54], Lese

and Banerjee [55], Moriguti [57], and Bhaskararao [23]). As we
proceed, we shall see what contributions these authors have made
in resolving some of the issues.

However, in order to understand the nature of their contributions and to comprehend their line of thinking, it would be
worthwhile to have some idea of how the efficiency of a weighing design has been defined by different workers at different
times. The various definitions are outlined in the following
section.

1.8   EFFICIENCY OF A WEIGHING DESIGN

From the way the weighing problem was originally posed by
Hotelling, it appeared that a weighing design could be taken as
the most efficient if each estimated weight had the minimum possible variance. But, because all the variances of the estimated
weights might not be equal in a design, one could call a weighing design the best, consistent with the above criterion, if the
average variance were the minimum. (This criterion is referred
to as A-optimality, see Kiefer [51, 52].)

Led by the consideration that the least possible variance
(in a chemical balance) of an estimated weight is σ^2/N, Kishen
[53] defined the efficiency of a given design as

$$\frac{\sigma^2/N}{\sigma^2 \, \Sigma_{i=1}^{p} \, c_{i1}/p} = \frac{p}{N \, \Sigma_{i=1}^{p} \, c_{ii}}$$

When, however, $N = p$, the above expression reduces to $1/\Sigma_{i=1}^{N} \, c_{ii}$.
This definition conveys the full idea of the efficiency of a
given weighing design in the chemical balance problem, in which
an estimated weight might have the least possible variance σ^2/N.
For instance, by this definition, an 8×8 design matrix in a

chemical balance problem will have 100% efficiency, as it should.
But this definition does not carry the full import of the efficiency of a given design in a spring balance problem, because the
minimum possible variance of σ^2/N is not attainable with a
spring balance. For example, in the spring balance design for
$N = p = 3$, design (1.4), which has the maximum possible efficiency, will not have 100% efficiency on this index. In spite
of this limitation, the relative efficiency of any two given designs can, in any case, be computed on the basis of this criterion, whether the balance is a chemical balance or a spring balance.

Mood [56] called a design the best if the determinant $|c_{ij}|$
$= \det|C|$ is minimized or, equivalently, if $\det|X'X| = \det|A|$
is maximized. (This criterion is referred to as D-optimality;
see Kiefer [51, 52].) A best design in this sense is therefore
a design that gives the smallest confidence region in the $\hat{\beta}_i$
$(i = 1, 2, ..., p)$ space for the estimates of the weights.[*]
On the basis of this formulation, the efficiency of a weighing
design can be defined [6] as follows:

Definition. Denote by c the minimum possible value of the
determinant $|c_{ij}|$. Then the ratio $c/|c_{ij}|$ furnishes the mea-
sure of efficiency of the design under consideration.

This definition can also be written as $|a_{ij}|/A$, where
A is the maximum possible value of $\det|A| = \det|X'X|$. In
this volume, we shall refer to this definition as Mood's "efficiency definition." By this definition, it is possible to calculate the efficiency of a given design when the value of A

---

[*]This definition is similar to Wald's generalized definition [70] of the measure of efficiency of a design for testing
a linear hypothesis in statistical investigations.

is known. In any case, however, the relative efficiency of any two given designs (with the same number of weighings) can be computed even though the maximum possible value of $\det|A|$ is not known.

Mood [56] pointed out further that other definitions of "best designs" might conceivably be preferred. Problems might arise in which one might wish to

1. minimize the variance factors subject to the restriction that they be equal,
2. minimize some function of the variance factors, or
3. minimize only a certain subset of the c_{ii} on a minor of the matrix (c_{ij}).

This might be the case if one wanted only rough estimates of the weights of some of the objects, but accurate estimates of the others.

Ehrenfeld [37] suggested a definition of efficiency for experimental designs in general, which might as well be applied to weighing designs. His definition is based on the maximization of λ_{min} (minimum eigenvalue of $[X'X]$). (Such criteria are referred to as E-optimality; see Kiefer [51, 52].)

Definition. Denote by u the maximum value of λ_{min} of $[X'X]$ for all x_{ij} in the relevant space. The ratio of λ_{min}/u is called the efficiency of a given design.

A design is called the most efficient if the efficiency of the design is equal to 1. (Wald [70] also visualized the possible use of such a definition.)

The above definition is referred to as Ehrenfeld's "efficiency definition. Following Ehrenfeld's efficiency definition, Raghavarao [61] suggested λ_{min}/N as the measure of efficiency of a given weighing design.

It has, however, been indicated [14] that, in defining the efficiency of a chemical balance design, the maximization of $|A|$ would be equivalent to the maximization of λ_{min}, whereas in the case of a spring balance design the criterion of maximization of $\det|A|$ would be preferable in that this criterion would include the consideration of maximization of λ_{min}. Thus, in defining the efficiency of a weighing design, a unified approach is provided by Mood's efficiency definition.

In some cases, the above criteria have been adopted as they are, without any restriction. In other cases, however, these criteria have been adopted subject to the restraints that (a) the variances of the estimated weights be equal and (b) the covariances between pairs of the estimated weights also be equal.

Chapter 2

CHEMICAL BALANCE PROBLEM

2.1   CHEMICAL BALANCE DESIGNS AND HADAMARD MATRICES

From the details given in Chapter 1, it is evident that if
N weighing operations are made to determine the weights of N
objects, the minimum variance that each estimated weight might
have is σ^2/N, and this minimum variance is reached when the design matrix X is orthogonal (orthogonal in the sense that
$[X'X]$ is diagonal), consisting entirely of +1 and -1 elements.
Thus, the problem of finding the best chemical balance design
is related, as pointed out by Mood [56], to Hadamard matrices
and the Hadamard determinant problem.

Hadamard proved the following result: If the elements x_{ij}
of a square matrix X are restricted to the range $-1 \leq x_{ij} \leq 1$,
the maximum possible value of the determinant of X is $N^{N/2}$
and, when this maximum value is achieved, all $x_{ij} = \pm 1$. The
matrix is then orthogonal in the sense that $[X'X]$ is diagonal.
The nonzero elements of $[X'X]$ are all equal to N.

We denote such a matrix by H_N. If H_N exists for a given
N, H_N is the best chemical balance design for $N = p$. If the

17

number of objects to be weighed is less than N, we may choose
from H_N a number of columns equal to the number of objects to
be weighed.

When H_N is used as the design matrix to determine the
weights of N objects, no degrees of freedom are left to estimate the error variance. On the other hand, if p columns of
H_N (p < N) are used as the design matrix to determine the
weights of p objects, the error variance can be estimated on
N - p degrees of freedom. If a larger number of degrees of
freedom is needed to estimate the error variance from the experiment itself, an H_N can be replicated as many times as needed.

With regard to the existence of H_N, it is known that a
necessary condition is

$N \equiv 0 \pmod{4}$

with the exception of N = 2. It is not known, however, whether
this condition is sufficient.

Kishen [53] pointed out that an H_N can easily be constructed when $N = 2^{2m}$ (for m, a positive integer) and gave
an illustration of an H_N when $N = 2^4 = 16$. In fact, such an
H_N (when N is a power of 2) is easily constructed by taking
the direct product of

$$H_2 = \begin{bmatrix} 1 & 1 \\ 1 & -1 \end{bmatrix}$$

For example,

$$H_4 = H_2 \otimes H_2 = \begin{bmatrix} 1 & 1 & 1 & 1 \\ 1 & -1 & 1 & -1 \\ 1 & 1 & -1 & -1 \\ 1 & -1 & -1 & 1 \end{bmatrix}$$

By continuing the above process m times, an H_N can be constructed when $N = 2^m$.

However, in attempting the construction of Hadamard matrices,
the author [5] indicated its connection with symmetrical balanced
incomplete block designs (BIBD). He observed [5] that one of
Bose's [24] methods of constructing symmetrical BIBD's could be
utilized in constructing a certain class of Hadamard matrices.
But none of these methods gave a comprehensive procedure for
constructing Hadamard matrices in general.

Mood [56] referred to the work of Paley [58] and Williamson
[71] and pointed out that H_{4k} exists for the range of

$$0 < 4k \leq 100$$

with the possible exception of $4k = 92$.

At about the time Mood's article [56] appeared, Plackett
and Burman [59] also provided solutions of Hadamard matrices for
all possible $N \leq 100$ except for $N = 4k = 92$. The Hadamard
matrix for $4k = 92$ remained undecided until 1962. The solution was first given by Baumert et al. [20]. (A comprehensive
account of Hadamard matrices is now available [44].)

2.2   CHEMICAL BALANCE WEIGHING DESIGNS WHEN N $\neq$ p

Kishen [53] pointed out that, with $N = 2^m + 1$ weighing
operations, the most efficient chemical balance weighing design
would probably be obtained by augmenting an H_N with a row,
1, 1, ..., 1. He also suggested that, with $N = 2^m + r \ (r < 2^m)$
weighing operations, a highly efficient design would be available if H_N were augment with r rows of +1's. With reference
to Kishen's [53] latter observation, Mood [56] pointed out that,
even with r = 2, it would be preferable to add two different
rows of H_N to H_N than to add two rows of +1's.

These observations clearly indicated the necessity for a detailed examination of the efficiencies of design matrices obtained by augmenting H_N with its rows, which are not necessarily the same. A comprehensive account of design matrices obtained by augmentations is available [4].

A word of caution is necessary regarding the meaning of the symbol N. It may be recalled that N was used in the model to denote the number of weighings, that is, the number of rows of X. Again, N of H_N denotes the dimensions of H_N. This N of H_N may not, however, be the same as the number of weighings. For example, if two rows of an H_N are added to an H_N, the number of weighings will be N + 2. What N means in terms of the number of weighings will, of course, be clear from the context.

A few of the variance factors obtained under different situations are given below [4]. (Here, N refers to the N of H_N.)

1. When N + 1 weighings are needed, any row of +1's or any row of +1's and -1's of an H_N may be added to an H_N. In any case, the resultant design matrix gives the same magnitude to the variance factors, and we have

$$c_{ii} = \frac{N + p - 1}{N(N + p)} \quad i = 1, 2, \ldots, p$$

2. When two different rows of an H_N are added to an H_N, two different sets of variance factors are obtained. If p is even and equal to r + s, each in the first set of r variance factors is equal to

$$\frac{N + 2r - 2}{N(N + 2r)} \tag{2.1}$$

and each in the second set of s variance factors is equal to

$$\frac{N + 2s - 2}{N(N + 2s)} \qquad\qquad\qquad (2.2)$$

In the above, we have

(a) $(2.1) < (2.2)$ if $r < s$

(b) When p is even, the minimum average variance is obtained when $r = s = p/2$. In this case, each c_{ii} is equal to

$$\frac{N + p - 2}{N(N + p)} \qquad i = 1, 2, \ldots, p$$

(c) When p is odd, the minimum average variance is obtained when $p = 2s + 1$. Each in the first set of $s + 1$ variance factors is equal to $(N + p - 1)/N(N + p + 1)$, and each in the second set of s variance factors is equal to $(N + p - 3)/N(N + p - 1)$.

(d) When, however, the same row is added twice, the variance factors, as shown by Mood [56], are

$$\frac{N + 2p - 2}{N(N + 2p)} \qquad\qquad\qquad (2.3)$$

(e) Each of (2.1) and (2.2) is less than (2.3) since $r, s < p$.

3. When $N + 3$ weighings are needed, three different rows of an H_N may be added to an H_N.

Different situations arise, depending on p and the number of different rows that are added to an H_N. One can obtain $N + 3$ weighings by deleting one row from H_{N+4}. These different cases have been studied and compared by the author [4].

2.3   CHEMICAL BALANCE DESIGNS FOR VALUES OF N = p
           WHEN H_N DOES NOT EXIST

2.3.1   Design Matrices When det$|A|$ Is Maximum

The following are some of the most efficient chemical balance designs furnished by Mood [56] for small values of N = p
when H_N does not exist. Here, a design is called the most
efficient (Mood's efficiency definition) if the value of det$|A|$
= det$|X'X|$ is maximized or, equivalently, if det$|C|$ is mini-
mized. Mood [56] constructed these square matrices following a
method of Williamson [72]. (The interested reader may also refer, in this connection, to the work of Gordon [41].)

There may be several designs with the same maximum value
for det$|A|$. When p = 3, the best designs (omitting 1 follow-
ing the sign) are

$$X = \begin{bmatrix} + & + & 0 \\ + & - & + \\ - & + & + \end{bmatrix} \qquad X = \begin{bmatrix} + & + & + \\ + & - & + \\ - & + & + \end{bmatrix} \qquad X = \begin{bmatrix} + & + & - \\ + & - & + \\ - & + & + \end{bmatrix}$$

All of these matrices have det$|A|$ = 16 (which is considerably
smaller than the value of 27 that det$|A|$ would have if an
optimum design with an orthogonal matrix existed). The first
of the above designs, for p = 3, gives

$$\{c_{ii}\} = \{3/8,\ 3/8,\ 1/2\}$$

whereas the second and third give

$$\{c_{ii}\} = \{1/2,\ 1/2,\ 1/2\}$$

For N = p = 5, the two best designs are

$$X = \begin{bmatrix} + & + & + & + & - \\ + & + & + & - & + \\ + & + & - & + & + \\ + & - & + & + & + \\ - & + & + & + & + \end{bmatrix} \quad \text{and} \quad X = \begin{bmatrix} + & - & - & - & - \\ + & + & + & - & - \\ + & - & + & - & + \\ + & - & + & + & - \\ + & + & - & + & + \end{bmatrix}$$

both of which have

$$\det |A| = 3^2 2^8 \quad \text{and} \quad \{c_{ii}\} = \{2/9, \; 2/9, \; 2/9, \; 2/9, \; 2/9\}$$

For $N = p = 6$, the best design is

$$\begin{bmatrix} + & - & - & - & - & - \\ + & - & - & - & + & + \\ + & - & - & + & + & - \\ + & - & + & + & - & + \\ + & + & + & - & + & - \\ + & + & - & + & - & + \end{bmatrix} \tag{2.4}$$

which has

$$\det |A| = 5^2 2^{10} \quad \text{and} \quad \{c_{ii}\} = \{1/5, \; \ldots, \; 1/5\}$$

It is not known to the author if any systematic attempt has been made to construct such square matrices with a maximized determinant for larger values of $N = p$.

2.3.2   Other Designs of Comparable Efficiency

A method by Plackett and Burman [59] for constructing Hadamard matrices was used by the author [5] to illustrate the construction of a 6×6 and a 10×10 orthogonal matrix consisting of the elements +1, -1 and 0. These matrices are reproduced below:

For $N = p = 6$,

$$X = \begin{bmatrix} 0 & + & + & + & + & + \\ + & 0 & + & - & + & - \\ + & + & 0 & + & - & - \\ + & - & + & 0 & - & + \\ + & + & - & - & 0 & + \\ + & - & - & + & + & 0 \end{bmatrix} \qquad (2.5)$$

(Reference is made to matrix (2.5) later in the chapter.)

For $N = p = 10$,

$$X = \begin{bmatrix} 0 & + & + & + & + & + & + & + & + & + \\ + & 0 & + & - & + & - & + & - & + & - \\ + & + & 0 & + & - & + & + & - & - & - \\ + & - & + & 0 & - & + & - & - & + & + \\ + & + & - & - & 0 & + & - & + & + & - \\ + & - & + & + & + & 0 & - & + & - & - \\ + & + & + & - & - & - & 0 & + & - & + \\ + & - & - & - & + & + & + & 0 & - & + \\ + & + & - & + & + & - & - & - & 0 & + \\ + & - & - & + & - & - & + & + & + & 0 \end{bmatrix} \qquad (2.6)$$

For (2.5), the variance factors are $c_{ii} = 1/5$. These are
the same in magnitude as those under the 6×6 design matrix of
maximized determinant given by Mood [56]. For (2.6), the variance factors are $c_{ii} = 1/9$. (It should be mentioned that such
orthogonal matrices can be constructed [34] for any N, where
$N = p^n + 1$, $p^n = 4t + 1$.)

Following the notation used by Raghavarao [60, 61], we denote the above series of design matrices by S_N. Raghavarao
[60, 61] constructed two series of weighing design matrices for
the cases when (a) N is odd and (b) $N \equiv 2$ (mod 4). In the
construction of these matrices, a design was considered to be

the best when $1/\Sigma_{i=1}^{N} c_{ii}$ was the maximum, subject, however,
to the restraints that (a) the variances of the estimated
weights be equal and (b) the estimated weights be equally
correlated.

If the above two restraints are to be met, [X'X] has to
be of the form

$$
\begin{bmatrix}
r & \lambda & \cdots & \lambda \\
\lambda & r & \cdots & \lambda \\
\cdots & \cdots & \cdots & \cdots \\
\cdots & \cdots & \cdots & \cdots \\
\lambda & \lambda & \cdots & r
\end{bmatrix}
\tag{2.7}
$$

This matrix is of dimensions $p \times p$ with r in the diagonal,
and λ elsewhere. We denote such a matrix by $M(r, \lambda)$. We
know that $\det|M(r, \lambda)| = (r - \lambda)^{p-1}[r + \lambda(p - 1)]$ and that
the inverse of $M(r, \lambda)$ is of a similar form given by

$$M^{-1}(r, \lambda) = M(r^*, \lambda^*)$$

where

$$r^* = \frac{r + \lambda(p - 2)}{(r - \lambda) [r + \lambda(p - 1)]}$$

and

$$\lambda^* = \frac{-\lambda}{(r - \lambda) [r + \lambda(p - 1)]}$$

When $N = p$, we have

$$1/\Sigma_1^N c_{ii} = \frac{(r - \lambda) [r + \lambda(N - 1)]}{N[r + \lambda(N - 2)]}$$

Maximizing the measure of efficiency as given above,
Raghavarao [60] showed that, when N is odd, [X'X] of the
best design, when it exists, has to be of the form $M(N, 1)$,

and that, when $N \equiv 2 \pmod 4$, $[X'X]$ of the best design, when
it exists has to be of the form $M(N - 1, 0)$. The first series
of these matrices is denoted by P_N, and the second is denoted
by S_N. For each of these series, Raghavarao [60] provided
illustrations, of which a P_5 and an S_6 are given below:

$$P_5 = \begin{bmatrix} - & + & + & + & + \\ + & - & + & + & + \\ + & + & - & + & + \\ + & + & + & - & + \\ + & + & + & + & - \end{bmatrix}$$

In this design, $\det|A| = 3^2 2^3$, and $\{c_{ii}\} = \{2/9, 2/9, \ldots, 2/9\}$.

$$S_6 = \begin{bmatrix} 0 & + & + & + & + & + \\ + & 0 & + & - & - & + \\ + & + & 0 & + & - & - \\ + & - & + & 0 & + & - \\ + & - & - & + & 0 & + \\ + & + & - & - & + & 0 \end{bmatrix} \tag{2.8}$$

For this design, $\det|A| = 5^6$, and $\{c_{ii}\} = \{1/5, 1/5, \ldots, 1/5\}$.

Raghavarao [60] demonstrated further that, subject to the
restraints that the individual variances be equal and that the
covariances between pairs be equal, that is, subject to $[X'X]$
being of the form $M(r, \lambda)$, P_N matrices are also the best weighing designs under Wald's and Ehrenfeld's efficiency definitions.
In P_N, N is odd. When $N \equiv 2 \pmod 4$, S_N matrices are also
the best weighing designs under Ehrenfeld's efficiency definition.

Again, when $N \equiv 2 \pmod 4$, the best weighing design matrix
X under Mood's efficiency definition is that for which $[X'X]$
$= M(N, 2)$. We denote such a series by T_N.

The best weighing design for $N = 6$ satisfying Mood's
efficiency definition, subject to the above restraints, is given
by [61]

$$T_6 = \begin{bmatrix} + & - & - & - & - & - \\ + & - & + & + & + & + \\ + & + & - & + & + & + \\ + & + & + & - & + & + \\ + & + & + & + & - & + \\ + & + & + & + & + & - \end{bmatrix} \qquad (2.9)$$

In this design, $\det|A| = 2^{14}$, and $\{c_{ii}\} = \{7/32, 7/32, \ldots, 7/32\}$.
The covariance between each pair of the estimated weights is
$-\sigma^2/32$.

2.4   COMPLEXITIES ARISING FROM DIFFERENT DEFINITIONS OF EFFICIENCY

2.4.1   Comparison of Efficiencies of Square Matrices

It is clear by now that different authors have chosen different criteria to define the efficiency of a weighing design.
A weighing design that is considered to be the best by one criterion may not be considered so by another. One might thus wish
to compare these designs with reference to such desirable criteria as the magnitudes of the variance factors and the value
of $\det|A|$. Such a comparison is possible and can now be pre-
sented, since we have several different designs for $N = p = 6$
which have been considered to be the best under one criterion
or another.

Design (2.4) (which we denote by W_6) was given by Mood
[56]. Determinant $|A|$ is the maximum, being equal to $5^2 2^{10}$,
and each c_{ii} is $1/5$. The covariances are not equal. An S_6,

that is, design (2.8) given by the author [5] and later by
Raghavarao [60], gives 1/5 to each c_{ii}. The covariances be-
tween different pairs are equal (zero). The value of $\det|A|$
is 5^6, which is less than $5^2 2^{10}$. A T_6, that is, design matrix
(2.9) of Raghavarao [61], gives to each c_{ii} the value 7/32,
which is greater than 1/5. The covariances are the same, being
equal to -1/32. The value of $\det|A|$ for T_6 is 2^{14}, which
is greater than 5^6, but less than $5^2 2^{10}$.

These results are summarized in Table 2.1.

TABLE 2.1

Comparison of Efficiencies of Square Matrices

Design	Suggested by	Value of $\det|A|$	Variance factors	Covariance between pairs
W_6	Mood	$5^2 2^{10}$	1/5	Not equal
S_6	Banerjee, Raghavarao	5^6	1/5	Equal
T_6	Raghavarao	2^{14}	7/32	Equal

2.4.2   Comparison of Efficiencies of Rectangular Matrices

We shall presently see that further complexities arise when
we compare rectangular matrices.

The designs discussed in the previous section (square
matrices) give the weights of six objects with six weighing
operations. Such designs do not leave any degrees of freedom
for estimating the error variance. In order that the variances
be estimated from the experiment itself, the number of objects
has to be less than the number of weighing operations. For
example, to determine the weights of four objects with six weighing operations, one might think of choosing any four columns of

W_6, S_6, or T_6. Other designs might also be adopted for such a purpose. For instance, if the weights of four objects were to be determined from six weighing operations, a comparable design would be given by

$$X = \begin{bmatrix} + & + & - & - \\ + & + & + & + \\ \hline + & - & + & - \\ + & - & - & + \\ + & + & - & - \\ + & + & + & + \end{bmatrix} = V_6 \quad (\text{say})$$

This design was obtained [4] by augmenting an H_4 with two different rows of the H_4. For this design, $\{c_{ii}\} = \{3/16, 3/16, \ldots, 3/16\}$. The objects fall into two groups with equal covariance between pairs in each group. The value of $\det|X'X|$ is equal to $\det|A| = 2^{10}$.

On the other hand, if any four columns of an S_6 are chosen, $\{c_{ii}\} = \{1/5, 1/5, \ldots, 1/5\}$, and the value of $\det|A|$ is 5^4. The covariances are the same, each being equal to zero.

If the first two and the last two columns of W_6 are adopted as the design matrix, the resultant matrix is identical with V_6. If the first three columns and the fifth column of W_6 are chosen, the resultant design matrix leads to two sets of variance factors. One set of three has the value $2/5$, and the other has the value $1/6$. The average is $41/120$, which is larger than $1/5$. The value of $\det|A|$ is $3^2 2^9$, which is larger than 2^{10}. (Note that whichever four columns of W_6 are chosen, the resultant design matrix is one of these two types. The covariances divide into two groups with equal magnitude in each group.)

If any four columns of a T_6 are taken as the design matrix, the variance factors are $5/24$, which is larger than $1/5$.

The value of $\det|A|$ is $3 \cdot 2^8$, and the covariances are equal. In this case, the value of $\det|A|$ is larger than that of an S_6.

The above results are summarized in Table 2.2.

Thus, if smaller magnitude of the variance factors or larger value of $\det|A|$ is taken as the acceptable criterion (sacrificing equality of covariances), V_6 is better than S_6 and T_6. On the other hand, if value of $\det|A|$ is taken as the acceptable criterion, T_6 is better than S_6. If magnitude of the variance factors is taken as the criterion, S_6 is better than T_6.

It can be further seen that V_6 is comparable to W_6, although the construction of a V_6 is relatively simpler [3], when an H_4 is known.

The value of $\det|A|$, for any V_{N+2} where $N \equiv 0 \pmod{4}$ and H_N exists, is $2^2 N^N$. If we adopt N columns of an S_{N+2} as the design matrix, the value of $\det|A|$ is $(N + 1)^N$. If we adopt N columns of a T_{N+2} as the design matrix, the value of $\det|A|$ is $3N^N$. Thus, the efficiency of a T_{N+2} relative to that of a V_{N+2}, in terms of the value of $\det|A|$ (that is, the ratio of the value of $\det|A|$ for T_{N+2} to that of V_{N+2}), is $3/4$. The efficiency of an S_{N+2} relative to that of a V_{N+2}, in terms of the value of $\det|A|$ is $[1 + (1/N)]^N/4$, which, however, approaches $e/4$, as $N \to \infty$. The variance factors in a V_{N+2} are, in general, less than those in an S_{N+2} and a T_{N+2}.

TABLE 2.2

Comparison of Efficiencies of Rectangular Matrices

Design	Suggested by	Value of $\det|A|$	Variance factors	Covariances between pairs
V_6	Banerjee	2^{10}	3/16	Fall into two groups with equal covariances in each group
S_6	Banerjee, Raghavarao	5^4	1/5	Zero (equal)
W_6 (if the first two and last two columns are chosen)	Mood	2^{10}	3/16	Fall into two groups with equal covariances in each group
W_6 (if four columns are arbitrarily chosen)	Mood	$3^2 2^9$	Avg, 41/120	
T_6	Raghavarao	$3 \cdot 2^8$	5/24	Equal

Chapter 3

SPRING BALANCE PROBLEM

3.1 TWO THEOREMS OF MOOD GIVING THE BEST
SPRING BALANCE DESIGNS

As mentioned before, the spring balance problem is different from the chemical balance problem in that the elements x_{ij} of the design matrix X are restricted to the values 0 and 1, whereas in the chemical balance problem the elements of the design matrix can assume the values +1, -1, or 0.

For number of weighings $N > p$, Mood [56] used the following approach to obtain the best design (best according to Mood's efficiency definition) for a spring balance.

"Let P_r be a matrix whose rows are all the arrangements of r ones and $(p - r)$ zeros $(0 \leq r \leq p)$. (The symbol should also have a subscript p, but that is omitted because any specific value for p will always be clear from the context.) The matrix will have p columns and $\binom{p}{r}$ rows. Let X be a matrix made up of matrices P_r arranged in vertical order.

Let n_r be the number of times P_r is used in constructing X.
X is a weighing design for p objects and $N = \Sigma_r \, n_r \, \binom{p}{r}$
weighing operations."

Adopting these notations, Mood [56] proved the following
two theorems giving the best spring balance design.

Theorem 1. If $p = 2k - 1$, where k is a positive integer,
and if N contains the factor $\binom{p}{k}$, then $|a_{ij}|$ $(\det|A|)$ will
be maximized when $n_k = N/\binom{p}{k}$ and all other $n_r = 0$.

Theorem 2. If $p = 2k$, where k is a positive integer, and if
N contains the factors $\binom{p+1}{k+1}$, then $|a_{ij}|$ $(\det|A|)$ will be
maximized when $n_k = n_{k+1} = N/\binom{p+1}{k+1}$ and all other $n_r = 0$.

Mood [56] observed that, when p is odd, P_k is a design
that not only minimizes the confidence region (that is, the
value of $\det|C|$) for estimating the weights, but also mini-
mizes the individual variance factors. However, when p is
even, Mood [56] observed that the variance factors may not be
the minimum and surmised that the best design from the viewpoint
of minimum variance factors would be made up largely from P_k
and a small proportion from P_{k+1}.

3.2 SPRING BALANCE WEIGHING DESIGNS AND
                          BALANCED INCOMPLETE BLOCKS

Theorem 1 implies that if, for instance, the number of objects is 15, the total number of weighing operations has to be
as large as $\binom{15}{8}$ in order that the maximum possible precision
be reached. This would obviously require a very large number

of weighing operations. Even for a smaller number of objects,
say 7, we would have to make $\binom{7}{4}$ = 35 weighing operations in
order to secure the maximum possible efficiency. Similar argu-
ments apply to the case when p is even.

However, it has been pointed out [1] that spring balance
designs of equivalent efficiency but based on a smaller number
of weighing operations are available from the the arrangements
given by balanced incomplete block designs discussed in Fisher
and Yates [40]. (Incidentally, a BIBD is an arrangement of v
distinct objects into b blocks such that each block contains
exactly k distinct objects. Each object occurs in exactly r
different blocks and every pair of distinct objects occurs together in exactly λ blocks.) Such designs are extensively
used in agrobiological experiments and are characterized by five
parameters, v, b, r, k, and λ, connected by the identities

bk = vr   λ(v - 1) = r(k - 1)

In weighing designs, v takes the place of p, the number
of objects to be weighed, and b replaces N, the number of
weighings that can be made. Matrix [X'X] in this case takes
the same form as M(r, λ), that is, a matrix with r on the diagonal and λ elsewhere. The variance factor for each estimated weight [1], as mentioned before, is

$$\frac{r + \lambda(p - 2)}{(r - \lambda) \, [r + \lambda(p - 1)]}$$

where p is the number of objects to be weighed, and r and λ
have the same meanings as in the theory of BIBD's; that is, r
is the number of times each object is weighed, and λ is the
number of times each pair of objects is weighed together.

Although the minimum minimorum of σ^2/N can never be attained in a spring balance design, σ^2/N can still be held as
the standard to which the efficiency of a given design is

compared. The efficiency of the above design is then given [1]
by

$$\frac{(r - \lambda) \, [r + \lambda(p - 1)]}{N[r + \lambda(p - 2)]} \tag{3.1}$$

Replacing N by b, and p by v, to accord with the notation of weighing designs, the two identities (as known in the
theory of BIBD's) can be rewritten as

$$r = \frac{Nk}{p} \quad \lambda = \frac{r(k - 1)}{p - 1}$$

Substituting these values in (3.1), we obtain the efficiency factor in the form

$$\frac{k^2(p - k)}{p(pk - 2k + 1)} \tag{3.2}$$

where k is the number of plots per block, or the number of objects that can be weighed at a time.

If, instead of adopting repetitions of P_k, only $\binom{p}{k}$
weighings are made in all, the efficiency factor for such designs
(that is, designs with all possible combinations of k objects
at a time) is given by

$$\frac{(r - \lambda) \, [r + \lambda(v - 1)]}{b[r + \lambda(v - 2)]}$$

where

$$r = \binom{v - 1}{k - 1} \quad \lambda = \binom{v - 2}{k - 2} \quad \text{and} \quad b = \binom{v}{k}$$

On simplification, the above expression reduces to (3.2). Thus,
the efficiency of a BIBD as a weighing design is the same as that
of the designs P_k of Mood [56].

If, on the other hand, the efficiency is calculated on the
basis of Mood's efficiency definition (that is, a design being

considered the best where $\det|A|$ is the maximum), it can also
be shown [7] that a BIBD as a weighing design has the same efficiency as the P_k of Mood, if, of course, the comparison is
made on the same number of weighings.

It can be seen from the above that the efficiency of such
designs depends only on p, the number of objects whose weights
are to be determined, and on k, the number of such objects that
can be weighed at a time.

The advantage of these designs is that all weights are
estimated with equal precision and the covariances between pairs
are the same.

3.3 SERIES OF DESIGNS DENOTED L_N BY MOOD

Mood [56] has also constructed spring balance designs of
maximum efficiency when $N = p$ and $N \equiv 3 \pmod{4}$. He has
shown that in such a case the spring balance design of maximum
possible efficiency is given by H_{N+1}, if it exists. For ready
reference, the method of construction as given by Mood [56] is
described below.

Let K_{N+1} denote the matrix formed from H_{N+1} by adding
or subtracting the elements of the first row of H_{N+1} from the
corresponding elements of the other rows in such a way as to
make the first element of each of the remaining rows 0. Obviously,

$$|K_{N+1}| = \pm |H_{N+1}|$$

Except for the first row, the elements of K_{N+1} are 0
and ± 2, with the signs of the nonzero elements being the same
for elements in the same row. Let L_N be the matrix obtained

by omitting the first row and column of K_{N+1}, by changing all
nonzero elements to +1, and by permuting two rows, if necessary,
to make the determinant of L_N positive. Then,

$$|H_{N+1}| = 2^N |L_N|$$

It is clear from the above that, given L_N, one could re-
verse the procedure and determine an H_{N+1}. In the same manner,
there is a correspondence, in general, between square matrices
with elements +1 and square matrices of one order less with elements 0 and 1. The ratio of the values of the corresponding
determinants is always 2^N, if their determinants do not vanish.
Hence, the (0, 1) determinant has its maximum possible value
when its corresponding (+1) determinant has the maximum pos-
sible value. Thus, $\det|L_N|$ has the maximum value.

3.4 CONNECTION OF L_N WITH A SYMMETRICAL BIBD
OF A SPECIAL KIND

An 8×8 orthogonal matrix [matrix (1.3), for instance]
affords an L_7, as given by

$$L_7 = \begin{bmatrix} 0 & 0 & 0 & 1 & 1 & 1 & 1 \\ 0 & 1 & 1 & 0 & 0 & 1 & 1 \\ 0 & 1 & 1 & 1 & 1 & 0 & 0 \\ 1 & 0 & 1 & 0 & 1 & 0 & 1 \\ 1 & 0 & 1 & 1 & 0 & 1 & 0 \\ 1 & 1 & 0 & 0 & 1 & 1 & 0 \\ 1 & 1 & 0 & 1 & 0 & 0 & 1 \end{bmatrix}$$

This L_7 is the same as L_7 of Mood [56] with a rearrangement
of its rows. As has been pointed out [1], L_7 is easily

recognized as a symmetrical BIBD for $v = b = 7$, $r = k = 4$, and
$\lambda = 2$. Such a design is characterized by an orthogonal series
[40]. The distribution of 1's shows the combination in which
the objects have to be weighed.

From the above it is clear that Hadamard matrices are connected with a special kind of symmetrical BIBD. For example,
an H_{16} leads to the design with $v = b = 15$, $r = k = 8$, and
$\lambda = 4$. In general, we have $v = b = N$, $r = k = (N + 1)/2$, and
$\lambda = (N + 1)/4$. With these values substituted for r and λ,
we have

$$\det|L_N| = \frac{(N + 1)^{(N+1)/2}}{2^N}$$

The variance factors are $c_{ii} = 4N/(N + 1)^2$.

Determinant $|L_N|$ and the variance factors can be easily
calculated in the above forms when it is recalled that the design matrix X (that is, L_N) is such that $[X'X]$ has the form
$M(r, \lambda)$.

Design (1.4), which Hotelling [49] indicated as the best
design for $N = p = 3$, evidently belongs to the series L_N
$(N = 3)$.

In an L_N, not only is $\det|A|$ maximized, but the vari-
ance factors are also minimized. That the variance factors are
minimized has been shown by Mood [56]. However, it has been
shown [3] that there are no combinations of the values of the
parameters in a BIBD other than those in an L_N, for which the
variance factors reach the minimum value of $4N/(N + 1)^2$.

3.5 USE OF BIBD'S AS WEIGHING DESIGNS

When a BIBD is used as a weighing design, the solutions of
the normal equations giving the estimated weights (least squares)
can be reduced to a simple routine [9]. The estimated weights
are given by

$$\hat{\beta}_i = \frac{1}{r - \lambda}\left[Z_i - \frac{\lambda T}{r + \lambda(v - 1)}\right]$$

$$= \frac{1}{r - \lambda}\left(Z_i - \frac{\lambda t}{r}\right)$$

where Z_i $(i = 1, 2, \ldots, p)$ are the elements of the vector $X'\underline{Y}$, $T = \Sigma_{i=1}^{v} Z_i$, $t = \Sigma_{i=1}^{b} y_i$, $v = p$, and $b = N$.

In particular, when L_7 is used as the design, the estimates are given by

$$\hat{\beta}_i = \frac{Z_i - (t/2)}{2}$$

While suggesting the use of BIBD's as efficient spring balance weighing designs, it might be worthwhile to point out that, when a BIBD exists with parameters v, b, r, k, λ, a complementary BIBD also exists with parameters

$$v_0 = b, \; b_0 = b, \; r_0 = b - r, \; k_0 = v - k, \; \text{and} \; \lambda_0 = b - 2r + \lambda$$

In the complementary design, each block is replaced by its complement. For example, the distribution of 1's in L_7 gives the design with $v = b = 7$, $r = k = 4$, and $\lambda = 2$, while the distribution of 0's gives the complementary BIBD with parameters $v = b = 7$, $r = k = 3$, and $\lambda = 1$. In general, the distribution of 0's in an L_N gives the complementary design for which $r_0 = r - 1$, $k_0 = k - 1$, and $\lambda_0 = \lambda - 1$.

3.6 EFFICIENT SPRING BALANCE WEIGHING DESIGNS
FOR SMALL VALUES OF $N = p$

Mood [56] quoted from Williamson [71] some square matrices with maximum value for $\det|A|$ to furnish weighing designs of maximum possible efficiency for small values of $N = p$. These square matrices are denoted by D_p, where p shows the dimension. For example, we have for $N = p = 4$

$$D_4 = \begin{bmatrix} 1 & 1 & 1 & 0 \\ 1 & 1 & 0 & 1 \\ 1 & 0 & 1 & 1 \\ 0 & 1 & 1 & 1 \end{bmatrix} \quad \text{and} \quad \begin{bmatrix} 1 & 0 & 0 & 1 \\ 1 & 1 & 1 & 0 \\ 0 & 0 & 1 & 1 \\ 0 & 1 & 0 & 1 \end{bmatrix}$$

The value of $\det|A|$ is 9. For the first, variance factors are all 7/9 and, for the second, $\{c_{ii}\} = \{7/9,\ 7/9,\ 4/9\}$.

For $N = p = 5$, we have

$$D_5 = \begin{bmatrix} 0 & 0 & 0 & 1 & 1 \\ 0 & 0 & 1 & 1 & 0 \\ 0 & 1 & 1 & 0 & 1 \\ 1 & 1 & 0 & 1 & 0 \\ 1 & 0 & 1 & 0 & 1 \end{bmatrix}$$

The value of $\det|A|$ is 25, and the variance factors are $\{c_{ii}\}$ = $\{19/25,\ 19/25,\ 16/25,\ 11/25,\ 16/25\}$. For $N = p = 6$, we have

$$D_6 = \begin{bmatrix} 1 & 1 & 1 & 0 & 0 & 0 \\ 1 & 0 & 0 & 0 & 1 & 1 \\ 1 & 0 & 0 & 1 & 1 & 0 \\ 0 & 0 & 1 & 1 & 0 & 1 \\ 0 & 1 & 1 & 0 & 1 & 0 \\ 0 & 1 & 0 & 1 & 0 & 1 \end{bmatrix}$$

The value of $\det|A|$ is 81, and the variance factors are $\{c_{ii}\}$ = $\{17/27,\ 17/27,\ 17/27,\ 17/27,\ 17/27,\ 17/27\}$.

It is not known to the author whether such square matrices have been constructed for higher dimensions in general.

3.7 SPRING BALANCE DESIGNS WHEN DIFFERENT ROWS
OF L_N ARE ADDED TO L_N

We have seen that H_N is the best chemical balance design for $N = p$ when it exists and that L_N is the best spring balance design for $N = p$ when H_{N+1} exists.

We noticed further that efficient chemical balance designs can be constructed between the dimensions of H_N and H_{N+4} by the addition of different rows of H_N to H_N and that some of these rectangular design matrices are comparable to chemical balance designs that are otherwise known to be the most efficient.

In the case of a spring balance, similar rectangular design matrices can be constructed between the dimensions of L_N and L_{N+4} by the addition of different rows of an L_N to the L_N.

When $N + 1$ weighings are required, any row of an L_N can be added to the L_N. When L_{N+2} weighings are required, either any row of an L_N can be added twice to the L_N or two different rows of an L_N can be added to the L_N. When $N + 3$ weighings are required, we can either add three rows of an L_N to the L_N or delete one row from L_{N+4}. A comparative study of the variance factors under these different situations has been made by the author [3].

3.7.1 A Useful Reduction of Symmetrical Determinant for Computational Purposes

In the study mentioned above, the following form of reduction of a symmetrical determinant [3] was found to be convenient for the computation of variance factors. Let

$$A_i = \begin{bmatrix} r_i & \lambda_i & \lambda_i & \cdots & \lambda_i \\ \lambda_i & r_i & \lambda_i & \cdots & \lambda_i \\ \vdots & & & & \\ \lambda_i & \lambda_i & \lambda_i & \cdots & r_i \end{bmatrix} \text{ and } \mu_{ij} = \begin{bmatrix} \lambda_{ij} & \lambda_{ij} & \cdots & \lambda_{ij} \\ \lambda_{ij} & \lambda_{ij} & \cdots & \lambda_{ij} \\ \vdots & & & \\ \lambda_{ij} & \lambda_{ij} & \cdots & \lambda_{ij} \end{bmatrix}$$

be, respectively, a square matrix of order N_i and a rectangular matrix of the same elements λ_{ij} having N_i rows and N_j columns. Also, let $\mu_{ji} = \mu'_{ij}$, the transpose of μ_{ij}.

In the calculation of variance factors, we often need to evaluate a determinant of the form

$$A_{1,2,\ldots,r} = \begin{vmatrix} A_1 & \mu_{12} & \mu_{13} & \cdots & \mu_{1r} \\ \mu_{21} & A_2 & \mu_{23} & \cdots & \mu_{2r} \\ \vdots & & & & \\ \mu_{r1} & \mu_{r2} & \mu_{r3} & \cdots & A_r \end{vmatrix} \qquad (3.3)$$

whose order is $N = \Sigma_{i=1}^{r} N_i$. The determinant has r symmetrical square matrices in the diagonal.

If $p_i = r_i + \lambda_i(N_i - 1)$, $P_i = (r_i - \lambda_i)^{N_i-1}$, and $p_i/N_i = R_i$, then determinant (3.3) reduces to

$$\prod_1^r P_i \prod_1^r N_i \begin{vmatrix} R_1 & \lambda_{12} & \lambda_{13} & \cdots & \lambda_{1r} \\ \lambda_{21} & R_2 & \lambda_{23} & \cdots & \lambda_{2r} \\ \vdots & & & & \\ \lambda_{r1} & \lambda_{r2} & \lambda_{r3} & \cdots & R_r \end{vmatrix}$$

where $\lambda_{ij} = \lambda_{ji}$. From the above it can be seen that the determinant of order $\Sigma_1^r N_i$ has been reduced to a determinant of order r, which may admit of a similar reduction for certain values of the elements.

There are r sets of variance factors and in each set
the variance factors are equal. Each of the N_i variance factors in the ith set is equal to

$$\frac{(N_i - 1)D'}{(r_i - \lambda_i)N_i D}$$

where

$$D' = \begin{vmatrix} R_1 & \lambda_{12} & \lambda_{13} & \cdots\cdots\cdots & \lambda_{1r} \\ \lambda_{21} & R_2 & \lambda_{23} & \cdots\cdots\cdots & \lambda_{2r} \\ \cdots & \cdots & \cdots & \cdots & \cdots \\ \lambda_{i1} & \lambda_{i2} & \lambda_{i3} & \cdots R_i' \cdots & \lambda_{ir} \\ \cdots & \cdots & \cdots & \cdots & \cdots \\ \lambda_{r1} & \lambda_{r2} & \lambda_{r3} & \cdots\cdots\cdots & R_r \end{vmatrix}$$

$$R_i' = \frac{r_i + \lambda_i(N_i - 2)}{N_i - 1}$$

and

$$D = \begin{vmatrix} R_1 & \lambda_{12} & \lambda_{13} & \cdots\cdots\cdots & \lambda_{1r} \\ \lambda_{21} & R_2 & \lambda_{23} & \cdots\cdots\cdots & \lambda_{2r} \\ \cdots & \cdots & \cdots & \cdots & \cdots \\ \lambda_{i1} & \lambda_{i2} & \lambda_{i3} & \cdots R_i \cdots & \lambda_{ir} \\ \cdots & \cdots & \cdots & \cdots & \cdots \\ \lambda_{r1} & \lambda_{r2} & \lambda_{r3} & \cdots\cdots\cdots & R_r \end{vmatrix}$$

Within the framework of this reduction, other simplifica-
tions can be introduced when convenient. To illustrate the use
of the reduction, let us refer to the design matrix D_6 given
by Mood [56]. If we rearrange the columns of D_6 so that the first
and fifth, the second and third, and the fourth and sixth columns are made to occupy consecutive positions, [X'X] for the
resultant design matrix takes the form

$$\begin{bmatrix} 3 & 2 & 1 & 1 & 1 & 1 \\ 2 & 3 & 1 & 1 & 1 & 1 \\ 1 & 1 & 3 & 2 & 1 & 1 \\ 1 & 1 & 2 & 3 & 1 & 1 \\ 1 & 1 & 1 & 1 & 3 & 2 \\ 1 & 1 & 1 & 1 & 2 & 3 \end{bmatrix} = A_{1 \cdot 2 \cdot 3} \quad (\text{say})$$

Then,

$$A_{1 \cdot 2 \cdot 3} = 1 \cdot 1 \cdot 1 \cdot 2 \cdot 2 \cdot 2 \cdot \begin{vmatrix} 5/2 & 1 & 1 \\ 1 & 5/2 & 1 \\ 1 & 1 & 5/2 \end{vmatrix} = 81$$

Thus, D of the above design matrix is equal to 81/8. D' (for any set of variance factors) would be of the form

$$\begin{vmatrix} 3 & 1 & 1 \\ 1 & 5/2 & 1 \\ 1 & 1 & 5/2 \end{vmatrix} = (3/2) \cdot 2 \begin{vmatrix} 3 & 1 \\ 1 & 7/4 \end{vmatrix} = \frac{51}{4}$$

Hence, the variance factors are

$$\frac{51}{4} \cdot \frac{8}{81} \cdot \frac{1}{1 \cdot 2} = \frac{17}{27}$$

3.7.2 Variance Factors When Different Rows of L_N Are Added to L_N

When a row of L_N is added t times to an L_N, the variance factors are the same, being equal to

$$\frac{4[N(t + 1) - t]}{(N + 1)^2 (t + 1)}$$

With $t = 1, 2, 3$, the variance factors reduce, respectively, to

$$\frac{2(2N - 1)}{(N + 1)^2} \tag{3.4}$$

$$\frac{4(3N - 2)}{3(N + 1)^2} \tag{3.5}$$

$$\frac{4N - 3}{(N + 1)^2} \tag{3.6}$$

When two different rows of an L_N are added to the L_N, the variance factors are again the same, being equal to

$$\frac{4(N - 1)}{(N + 1)^2} \tag{3.7}$$

Factors (3.7) are always less than (3.5) [3].

For $N + 3$ weighings, different situations arise, depending on how the rows of the design matrix are obtained. A detailed discussion of all such cases has been furnished by the author [3].

We can compare the variance factors obtained from a design given by three columns of a D_5 with those given by (3.7) when $N = 3$. If we need to make five weighing operations to determine the weights of three objects, it is better to add two different rows to L_3 than to use any three columns of D_5.

If a D_p of requisite dimensions is not available, we might choose an L_N of appropriate dimensions and add to it its rows to secure a substitute design.

It is not known to the author whether square matrices of the type D_p are available for larger values of p. There exists, of course, a need for a comprehensive compilation of such matrices.

3.8 CONSTRUCTION OF WEIGHING DESIGNS BY THE INCLUSION OF
ADDITIONAL ROWS TO A DESIGN MATRIX

It has already been pointed out that it is possible to
obtain the estimates of N weights with N weighings in a
chemical balance with maximum efficiency if the Hadamard matrix
H_N ($N = 4k$) exists. If the Hadamard matrix of the next higher
order (that is H_{N+4}) is available, it is possible to estimate
the weights of $N + 4$ objects with $N + 4$ weighings. No general solution, as far as is known to the author, is yet available by which it is possible to obtain a chemical balance design of maximum efficiency to estimate the weights of $N + m$
($m = 1, 2, 3$) objects with $N + m$ weighings when $N = 4K$.

It was indicated earlier that a spring balance design of
maximum efficiency to weigh $N - 1$ objects with $N - 1$ weighings is available if Hadamard matrix H_N exists. Here also,
no general solution appears to have been attempted for constructing a square matrix between the two most efficient spring balance designs of successive orders.

Williamson [72, 73], however, has outlined for small values of N a method of constructing determinants of order $N + m$
($m = 1, 2, 3$) with maximum possible values. The corresponding
matrices can be utilized as the most efficient weighing designs
(most efficient according to Mood's efficiency definition).

In the absence of the above matrices, there is some interest in determining the most efficient rectangular design
matrices of dimensions $(N + m) \times N$ ($m = 1, 2, 3$) in order to
estimate the weights of N objects with $N + m$ ($m = 1, 2, 3$)
weighings with the maximum possible efficiency. A procedure of
constructing such rectangular matrices has been outlined by the
author [7]. This was done by way of indicating a result which
may be called some sort of an extension of a result by Williamson [72, 73]. In the construction of rectangular matrices of

maximized efficiency, a design was considered to be the best
when $\det|A|$ was of the maximum possible value (Mood's effi-
ciency definition).

3.9 USE OF PARTIALLY BALANCED INCOMPLETE BLOCK DESIGNS
AS WEIGHING DESIGNS

We have seen that arrangements given by BIBD's can be
used as efficient spring balance designs. The most efficient
spring balance weighing designs L_N of Mood [56] are, as indicated earlier, given by a special class of symmetrical BIBD's.

It has been shown [10] that a general class of combinatorial arrangements developed by Bose and Nair [25] and known
as partially balanced incomplete block designs (PBIBD) may also
be used as weighing designs in some cases. But all such PBIBD's
cannot be used as weighing designs for estimating the weight of
each object uniquely in the least squares sense, because the
design matrix X might be singular in some cases. An indica-
tion has been given [10] of the conditions under which a design
matrix coming from a PBIBD may become deficient in rank. (Reference is made to a PBIBD in Sec. 4.6, where it is shown that if
the deficiency of the PBIBD is made up with a view to maximizing
the efficiency of the resultant design matrix, the resultant design matrix tends to become a BIBD.)

The D_6 quoted by Mood [56] for a spring balance weigh-
ing design for $N = p = 6$, maximizing $\det|A|$, is easily rec-
ognized [10] as a PBIBD with the following parameters:

$$v = b = 6, \; r = k = 3, \; \lambda_1 = 1, \; n_1 = 4, \; \lambda_2 = 2, \; n_2 = 1$$

$$p_{ij}^1 = \begin{bmatrix} 2 & 1 \\ 1 & 0 \end{bmatrix} \quad \text{and} \quad p_{ij}^2 = \begin{bmatrix} 4 & 0 \\ 0 & 0 \end{bmatrix}$$

If the first and fifth, the second and third, and the fourth and sixth columns of the above-mentioned D_6 are made to occupy contiguous positions, $[X'X]$ takes the form

$$\begin{bmatrix} 3 & 2 & 1 & 1 & 1 & 1 \\ 2 & 3 & 1 & 1 & 1 & 1 \\ 1 & 1 & 3 & 2 & 1 & 1 \\ 1 & 1 & 2 & 3 & 1 & 1 \\ 1 & 1 & 1 & 1 & 3 & 2 \\ 1 & 1 & 1 & 1 & 2 & 3 \end{bmatrix}$$

The variance factors for this design have already been provided.

A special class of PBIBD has been discussed in an article by the author [10], where the variance factors have been evaluated along with other details.

The observations in this section have been made to point out that we can also use PBIBD's as efficient spring balance weighing designs, just as we can use BIBD's for such purposes.

Chapter 4

MISCELLANEOUS ISSUES CONCERNING THE WEIGHING PROBLEM

4.1 DETERMINATION OF TOTAL WEIGHT

4.1.1 Efficient Designs for Estimating Total Weight

An orthogonal design that has the maximum efficiency in determining individually the weights of p objects in a chemical balance is not the best design for the estimation of a linear function of the objects. To illustrate this, let there be three objects, β_1, β_2, β_3, the weights of which have to be determined on a balance corrected for bias, and let us, for this purpose, adopt the following design:

$$X = \begin{bmatrix} 1 & 1 & 1 \\ 1 & 1 & -1 \\ 1 & -1 & 1 \\ 1 & -1 & -1 \end{bmatrix} \qquad (4.1)$$

In this design, the variance of each of the estimated weights is $\sigma^2/4$, which is the minimum minimorum, and, as such,

the design has the maximum efficiency in estimating the weights
of the individual objects. But for estimating a linear function
of the objects (for instance, the total weight) designs more
efficient than this are available.

The variance of $\ell_1\hat{\beta}_1 + \ell_2\hat{\beta}_2 + \ell_3\hat{\beta}_3$ is known to be

$$\sum_{i,j=1}^{3} \ell_i \ell_j c_{ij} \sigma^2 \tag{4.2}$$

where c_{ij} denotes the elements of $[X'X]^{-1}$. Since design
(4.1) furnishes the estimates orthogonally. The variance of
the estimated total weight is given by $3\sigma^2/4$.

If, instead, we adopt the design given by the matrix

$$X = \begin{bmatrix} 1 & 1 & 1 \\ 1 & 1 & 0 \\ 1 & 0 & 1 \\ 0 & 1 & 1 \end{bmatrix} \tag{4.3}$$

the variance of the estimated total weight can easily be seen
to be $3\sigma^2/7$, which is less than $3\sigma^2/4$. Therefore, with four
weighing operations, the design given by (4.3) is more efficient
in estimating the total weight than the design given by (4.1).
A still more efficient design for estimating the total weight
is simply to weigh all the objects together four times. The
necessity for an efficient design to estimate any linear function of the objects (or the total weight) will perhaps arise
only when the objects cannot all be weighed collectively on a
single pan.

In the estimation of total weight, an efficient design is
also afforded by the arrangements of a BIBD, because the covariances are all negative. It has been shown [1] that the variance
of the estimated total weight, using a BIBD, reduces to

$$\frac{p\sigma^2}{r + \lambda(p - 1)}$$

In practice, an object may break into pieces, and we may
need to know the total weight of the pieces (see Rao [64]).

4.1.2 Efficient Designs in Estimating Both Individual Weights and Total Weight

In the estimation of total weight, a BIBD, as a weighing
design, may sometimes be preferable to an optimum chemical balance design. But, a statement favoring a BIBD will not hold
good in general. We may have to construct special designs to
meet special needs. The question of constructing weighing designs that will be good both for "individual" estimation and for
"total" estimation was considered by the author [7]. In the
construction of such designs, the purpose was to minimize
$\Sigma_{i,j}^{p} c_{ij}$ under the conditions that

1. the variance of an estimated weight be equal to the
   variance of the total weight, and
2. the covariances between pairs be the same.

If condition 2 were to be satisfied, the form of [X'X] would
have to be that of M(r, λ).

In order that both 1 and 2 be met simultaneously, we would
have to proceed [7] in the following way:

Let us take the case of finding the weights of, say, eight
objects. An $H_N((N = 8)$ would furnish the optimum design for
the estimation of individual weights. If any of the eight rows
of H_N were added to it eight times, the corresponding [X'X]
would have r = N = 16 on the diagonal,and $\lambda = N/2 = 8$ elsewhere. It should be mentioned here that the addition of one
row of +1's and -1's would be equivalent, in terms of the variance factors, to the addition of a row of +1's. The addition

of a row of +1's would, however, imply placing all the objects
together in a single pan. On the other hand, if any row other
than a row of +1's were added, the net result in terms of the
variance factors would remain the same, but with the advantage
that we would not have to place all the objects together in one
pan.

By making 16 weighings, as indicated above, the precision
of both "individual" estimation and "total" estimation is increased up to the maximum point. The variance for each estimated individual weight and total weight is $\sigma^2/9$.

If, in the above design, the row is repeated 6 times instead of 8, making 14 weighings in all, the precision is necessarily lower. The variance of each estimated weight and that
of the total weight are $25\sigma^2/224$ and $\sigma^2/7$, respectively. It
is interesting to compare the efficiency of this design with
that given by a BIBD with the same number of weighings, since
a BIBD might ordinarily be thought to be a better design in
estimating total weight because it gives negative covariances.
A comparable BIBD is given by b = 14, v = 8, k = 4, r = 7, and
λ = 3. For this BIBD, the variances are, respectively, $25\sigma^2/112$
and $2\sigma^2/7$, which are double the values mentioned above. Therefore, in a chemical balance, when both pans can be used, the designs discussed here are more efficient than those afforded by
BIBD's for the purpose of estimating simultaneously the individual weights and the total weight.

4.2 FACTORIAL APPROACH TO THE WEIGHING PROBLEM

4.2.1 Fractional Replicates as Weighing Designs

Kempthorne [50] discussed the weighing problem from the
viewpoint of "factorial experimentation." He suggested that

the problem of weighing a number of objects be regarded as a
problem of estimating the effects of a number of factors that
do not interact. The motivation in this suggestion appears to
be in keeping with the tenor of the original example of Yates
[78], which was, in fact, meant to show that one could think of
a factorial experiment in which factors do not interact.

In his article [50], Kempthorne briefly described the
method of constructing fractional replicates of a 2^n-factorial
experiment, originally discussed by Finney [39], and showed how
a fractional replicate could be used as a weighing design to
find the weights of ten different objects, a, b, c, d, e, f, g,
h, k, ℓ. In the fractional replicate of 1 in 64 of the 2^{10} design, the identity relationship was taken to consist of all the
members of the Abelian group obtained from all combinations of
the elements (main effects and/or interactions), I, ABC, CDE,
EFG, GHK, ADL, and AFH. The actual design consisting of the 16
weighings is given by the following combinations:

(1)	abdef	acefℓ	bcdℓ	
abfgkℓ	degkℓ	bcegk	acdfgk	
fgh	abdegh	aceghℓ	bcdefghℓ	(4.4)
abhkℓ	defhkℓ	bcefhk	acdhk	

In this scheme, (1) corresponds to the "weighing operation" on
an empty pan, which, of course, is equivalent to determining
the bias. Each of the remaining 15 combinations represents a
weighing operation with the objects included in the combination.

The weight of a is one-eighth the difference between
those weighings containing a and those not containing a. The
weights of the other objects are found in a similar manner. In
the language of factorial experimentation, each of these estimates represents a "contrast." There are 10 such contrasts of
main effects which estimate the 10 weights.

The remaining five contrasts can be used to obtain an estimate of the experimental error. If σ^2 is the variance of each weighing, the variance of the estimated weight of a (that is, the main effect "A" in terms of the language of factorial experimentation) is $(1/8 + 1/8)\sigma^2 = \sigma^2/4$.

Kempthorne [50] mentioned further that the precision can be increased fourfold by interpreting the absence of each letter as the placing of the object in the other pan, in case a chemical balance can be used. This improvement is of the same nature as that indicated by Hotelling [49] with reference to the original illustration furnished by Yates [78].

Kempthorne [50] mentioned that such designs have the following useful properties:

1. The design automatically takes care of any bias in the balance.
2. The effects or the weights can be computed easily, as indicated above.
3. The effects are uncorrelated.
4. All the effects are measured with the same precision.
5. An estimate of the experimental error which is independent of the effects can be computed from the results.

These properties can be verified when the design matrix X corresponding to scheme (4.4) is written out in full and matrix $[X'X]^{-1}$ is evaluated.

4.2.2 Kempthorne's Observations About the Nonorthogonality of the Estimates Furnished by L_N of Mood

Of the properties listed in the previous section, property 3 has special significance. Kempthorne [50] mentioned that, although Mood's [56] optimum spring balance designs L_N furnish "somewhat" smaller variance than that given by the fractional

replicates, the optimum designs of Mood have the disadvantage
that the estimates are correlated, whereas estimates furnished
by fractional replicates are orthogonal.

It has, however, been shown [2] that the designs given by
fractional replicates suggested by Kempthorne [50] are virtually
the same as the designs L_N of Mood. The designs furnished by
fractional replicates take account of the bias, and if the weighing operation corresponding to the determination of the bias is
omitted (in case the spring balance is free from bias), the resultant design matrix will be the same as L_N of Mood and will
therefore fail to give orthogonal estimates. These optimum
designs (L_N of Mood) can also be made to furnish orthogonal
estimates when the designs are adjusted to suit estimation in
a biased balance. For example, it is true that the design matrix L_3 given by

$$X = \begin{bmatrix} 1 & 1 & 0 \\ 1 & 0 & 1 \\ 0 & 1 & 1 \end{bmatrix}$$

does not give orthogonal estimates as such. But let us assume
that the spring balance has a bias and that it is necessary to
find the estimates of weights free of bias. The design matrix
must then be modified to suit the required estimation. The
modification requires that one more column be added to the design matrix X to correspond to the estimation of the bias. If
the first column is taken to correspond to the bias and an additional weighing operation is made on an empty pan to determine
the bias, the modified design matrix is

$$X = \begin{bmatrix} 1 & 0 & 0 & 0 \\ 1 & 1 & 1 & 0 \\ 1 & 1 & 0 & 1 \\ 1 & 0 & 1 & 1 \end{bmatrix}$$

In terms of factorial experiments, this design can be regarded as a 1/2 replicate of a 2^3-factorial experiment being given by the identity relationship I = ABC. The combinations are (1, ab, ac, bc).

The first row of the above design represents a weighing operation on an empty pan and thus goes to the determination of the bias. The remaining three weighing operations represent weighings of the three objects taken two at a time along with the bias. This does not mean, however, that we need to make any fresh weighing operations. In fact, we use the same records of weighings that were available on the supposition that the balance had no bias. Only one additional weighing is needed, and this is a weighing on an empty pan. In substance, therefore, this modification really means that we need to make only one additional weighing operation, and this is done on an empty pan.

When the modification is made as above, we are able to determine the weights of the objects free of bias, and in mutually orthogonal linear combinations of the observations.

For the above modified design matrix, $[X'X]^{-1}$ takes the following form [2]:

$$[X'X]^{-1} = \begin{bmatrix} 1 & -1/2 & -1/2 & -1/2 \\ -1/2 & 1 & 0 & 0 \\ -1/2 & 0 & 1 & 0 \\ -1/2 & 0 & 0 & 1 \end{bmatrix}$$

It can be seen from the above that the variance of an estimated weight is σ^2, whereas the same for an L_3 is $3\sigma^2/4$.

All the rows of $[X'X]^{-1}X'$ except the first are orthogonal to one another, meaning that the estimates given by $\hat{\beta} = [X'X]^{-1}X'\underline{y}$ (except that for the bias) are linear functions of the y's which are mutually orthogonal. In fact, for this design, $[X'X]^{-1}X'$ reduces to

$$\begin{bmatrix} 1 & 0 & 0 & 0 \\ -1/2 & 1/2 & 1/2 & -1/2 \\ -1/2 & 1/2 & -1/2 & 1/2 \\ -1/2 & -1/2 & 1/2 & 1/2 \end{bmatrix}$$

The last three rows of the matrix show that the three estimates are obtained as orthogonal linear functions of the observations.

The above property holds for any N of L_N. Because L_N furnishes a special kind of symmetrical BIBD, it is natural to ask if this property of orthogonality of the estimates can also be extended to BIBD's in general. In fact, it can be, and this topic is discussed in Sec. 4.3. The deduction with respect to an L_N, as shown above, follows as a special case from the general development outlined in Sec. 4.3.

It was mentioned earlier that the distribution of 0's in L_N gives the complementary BIBD for which $r_0 = r - 1$, $k_0 = k - 1$, and $\lambda_0 = \lambda - 1$. If, to such a complementary design, a row of 1's and a column of 1's (in that order) are added to suit estimation in a biased spring balance, a similar situation is obtained [2], in that the estimates are mutually orthogonal. The variance factors of the estimated weights remain the same [8] as those obtained by modifying the original design L_n. It can readily be verified that the original design of Yates [78] to determine the weights of seven light objects and a bias is an illustration of this kind. If a column of 1's is added to the design of Yates [78], the resultant scheme is an 8×8 design matrix, and this design matrix is that which is obtained by the addition of a row of 1's and a column of 1's, in that order, to the design complementary to L_7.

One more observation is pertinent in this context. If, after the weighing operations are completed and the estimates are provided, it is detected that the balance has a bias, the records of weighings already obtained from an L_N are still

useful, but an additional weighing must be made on empty pan
(denoted y_0). After y_0 is obtained, the correction of $-y_0/d$
$(d = (N + 1)/2)$ is applied to all the estimates.

4.2.3 Use of Fractional Replicates of Other Types

While referring to the construction of optimum designs
for a chemical balance through fractional replicates, Kempthorne
mentioned the possibility of using a 3/4 replicate as a weighing design. It has been shown [2] that, when a 3/4 replicate
of a 2^n-factorial experiment is used as a chemical balance
weighing design, the variance factors are $1/2^{n-1}$. Again,
$1/2^{n-1}$ is the variance factor if the fractional replicate is
of the type $(2^\beta - 1)/2^\beta$ $(1 \leq \beta \leq n)$.

However, if a 3/4 replicate or a replicate of the type
$(2^\beta - 1)/2^\beta$ $(1 \leq \beta \leq n)$ is used as a spring balance weighing design, there are two sets of variance factors in each case
[2, 3]. Of $3 \cdot 2^{n-2}$ variance factors in a 3/4 replicate, a
set of three (the set corresponding to the defining contrasts)
has the value $3 - 1/2^{n-3}$, and the remaining $3(2^{n-2} - 1)$ variance factors have the value $1/2^{n-3}$. When the fractional replicate is of the type $(2^\beta - 1)/2^\beta$ $(1 \leq \beta \leq n)$, each of the
$(2^\beta - 1)$ variance factors in the first set is equal to

$$\frac{2^{\beta-1}(2^\beta - 1)t - 2^{\beta+1} + 4}{2^{n-1}(2^{\beta-1} - 1)^2}, \quad t = 2^{n-\beta}$$

and each of the remaining variance factors is equal to $1/2^{n-3}$.
Since $1/2^{n-1}$ is a quarter of $1/2^{n-3}$, this agrees with
the finding that spring balance designs may be at best one-
fourth as efficient as chemical balance designs.

4.3 HOW BALANCED INCOMPLETE BLOCK DESIGNS CAN
FURNISH ORTHOGONAL ESTIMATES
IN WEIGHING DESIGNS

4.3.1 Introduction

As mentioned in Sec. 4.2.2, Kempthorne [50] suggested
that the fractional replicate designs which afford uncorrelated
estimates of weights are preferable to the optimum designs L_N
of Mood [56], because in an L_N the estimates are correlated.
In response to this comment, it was pointed out [2] that the
designs L_N of Mood can also be made to furnish orthogonal
estimates (see Sec. 4.2.2). In fact, these two types of designs are virtually the same, since a slight modification in
one gives rise to the other.

We now intend to show that BIBD's in general can also be
made to furnish [8] uncorrelated estimates and that the solution in the case of an L_N is a particular case of the general
solution proposed here.

Writing the estimates as

$$\hat{\underline{\beta}} = [X'X]^{-1}X'\underline{Y} = B\underline{Y}$$

where $B = [X'X]^{-1}X'$, we have

$$[BB'] = ([X'X]^{-1}X')([X'X]^{-1}X')' = [X'X]^{-1}$$

The ith row of B shows how the y's are combined in the
ith estimate. Thus, in order that the estimated weights be
available in mutually orthogonal linear functions of the variables y_i $(i = 1, 2, \ldots, b)$, $[X'X]^{-1}$ has to be of diagonal
form. The constitution of $[X'X]^{-1}$ will therefore reveal which
of the estimates are mutually orthogonal.

4.3.2 Orthogonal Estimates Obtained When the Design Matrix Is
       Modified to Suit Estimation in a Biased Balance

If the balance has a bias, it is necessary to modify the
design matrix to find the estimates of weights free of bias.
This is achieved with an L_N in the following way: Taking the
bias as an additional object, the weight of which is to be
determined, a column of 1's and a row of 0's, in that order,
are added to the design matrix L_N. In BIBD's, in general, we
add a column of 1's and t rows of 0's in that order, to the
design matrix. This means devoting t weighings to the estimation of bias. It has been shown [8] that if t is chosen as

$$t = \frac{r^2}{\lambda} - b \qquad\qquad\qquad (4.5)$$

$[X'X]^{-1}$ reduces to

$$\begin{bmatrix} 1/t & -1/tk & -1/tk & \cdots & -1/tk \\ -1/tk & r/tk\lambda & 0 & \cdots & 0 \\ -1/tk & 0 & r/tk\lambda & \cdots & 0 \\ \cdots\cdots\cdots\cdots\cdots\cdots\cdots\cdots\cdots\cdots\cdots\cdots\cdots\cdots \\ -1/tk & 0 & 0 & & r/tk\lambda \end{bmatrix} \qquad (4.6)$$

Scheme (4.6) shows that, except for the bias, the other estimates are mutually orthogonal; $B = [X'X]^{-1}X'$ is available in
the form

$$B = \begin{bmatrix} 1/t & 1/t & 1/t & 0 & 0 & 0 & \cdots \\ -1/tk & -1/tk & -1/tk & 1/r & 1/r & 1/r & \cdots \\ -1/tk & -1/tk & -1/tk & 1/r & -1/tk & -1/tk & \cdots \\ \cdots\cdots\cdots\cdots\cdots\cdots\cdots\cdots\cdots\cdots\cdots\cdots\cdots\cdots \\ \cdots\cdots\cdots\cdots\cdots\cdots\cdots\cdots\cdots\cdots\cdots\cdots\cdots\cdots \end{bmatrix}$$

For every 0 and 1 in all rows of X' except the first, there
will be, respectively, -1/tk and 1/r in all rows of (4.6)
except the first. In every row except the first, there will be
$t + (b - r)$ 0's and r 1's. All rows of the above scheme are
mutually orthogonal, except the first. (Details of proof have
been furnished by the author [8].) It can easily be shown that
the sum of the elements in a row (except the first) is zero.
Thus, if the observation y is subject to a "constant" bias,
environmental or otherwise, the bias is cancelled out in the
estimates. This a noteworthy advantage.

Only those BIBD's for which an integral solution for t
in Eq. (4.5) exists would come under this class. That is, when
such an integral value of t exists, it is possible to obtain
orthogonal estimates. Orthogonal estimates are also available
from the BIBD's complementary to these BIBD's. In a complementary design, we have to add a column of 1's and t rows of 1's
in that order, to the design matrix. It has been indicated [8]
that the variance factors for the estimated weights in a complementary BIBD are the same as in the BIBD itself.

In an L_N, in particular, $r = (N + 1)/2$, $\lambda = (N + 1)/4$,
and $b = N$. Hence, $t = 1$. Thus, the procedure adopted in an
L_N is a special case.

The above procedure fails to furnish orthogonal estimates
when r^2/λ is not an integral number. A solution, however,
always exists for all BIBD's when $\lambda = 1$.

When r^2 is not divisible by λ, we assume that we can
find the least integer ξ such that $(r + \xi)^2$ is divisible
by $\lambda + \xi$. If a column of 1's and ξ rows of 1's and η rows
of 0's in that order, are added to the design matrix to suit
estimation in a biased balance, we obtain orthogonal estimates,
provided η is taken [8] as

$$\eta = \frac{(r + \xi)^2}{\lambda + \xi} - (b + \xi)$$

The above procedure also extends to complementary designs, in which we have to add η rows of 1's and ξ rows of 0's, rather than ξ rows of 1's and η rows of 0's. (Algebraic demonstrations of these results, along with numerical illustrations, have been furnished by the author [8]. The results reported by Lese and Banerjee [55] are also of interest in this context)

4.4 YATES' ORIGINAL EXAMPLE: THE BEST DESIGN FOR THE PROBLEM UNDER CONSIDERATION

4.4.1 A Critical Look at Yates' Original Example

Yates' original example [78] was a scheme that provided the estimates of seven light objects based on eight weighing operations undertaken with a <u>biased</u> spring balance (one-pan balance). Let the problem be posed in the following manner: Given a biased spring balance, what is the best design to provide the weights of seven light objects with the minimum of eight weighing operations? Since the balance is biased, eight weighing operations at the minimum have to be performed under the customary technique: one weighing for each of the seven objects and an additional weighing to determine the zero correction.

It has been shown [17] that Yates' original scheme was the best design for the problem under consideration. However, it has also been pointed out [17] that there exists a similar design with exactly the same properties of D-optimality, minimum variance, and orthogonality of the estimates.

The problem is approached through L_N, which is known to be the most efficient spring balance design to determine the weights of N objects with N weighing operations.

4.4.2 The Problem Approached Through L_N

When an L_N is used as a design in a biased spring balance problem, a column of 1's is automatically added to the design matrix. Without loss of generality, this column can be
taken to occupy the first position. Then, the resultant design
matrix is singular with deficiency in rank by 1. At least one
row must therefore be added to it to make up for this deficiency.
Let a row of $(1 + n)$ 1's $(n \le N)$ and $(N - n)$ 0's be added
to it, and let it occupy the first position. We find n so
that the resultant design is D-optimal. Matrix $[X'X]$ of the
resultant design is given by

$$[X'X] = \begin{bmatrix} (N+1)I_1 & (r+1)E_{1,n} & rE_{1,N-n} \\ (r+1)E_{n,1} & (r-\lambda)I_n+(\lambda+1)E_{n,n} & \lambda E_{n,N-n} \\ rE_{N-n,1} & \lambda E_{N-n,n} & (r-\lambda)I_{N-n}+\lambda E_{N-n,N-n} \end{bmatrix}$$

$$(4.7)$$

where $E_{a,b}$ is an a × b matrix with 1's as its elements, I_c is
an identity of order c, $r = (N + 1)/2$, and $\lambda = (N + 1)/4$.

The determinant of (4.7) is given by

$$|X'X| = \left(\frac{N + 1}{4}\right)^{N-1} \left(n - \frac{N + 1}{2}\right)^2 \quad n \le N \qquad (4.8)$$

The maximum of (4.8) is attained when n = 0 and is given by

$$(N + 1)^{N+1}/4^N \qquad (4.9)$$

Thus, in order that the augmented square design matrix be D-
optimal, n has to be 0.

Matrix (4.7), in the special case of n = 0, was obtained
and inverted by the author [8] in the context of providing

orthogonal estimates, but it was not indicated that the design
was D-optimal, because this was not the aim of the article [8].
The first diagonal element of the inverse is 1, and the other
diagonal elements are $4/(N + 1)$, implying that the variance of
the estimated bias is σ^2 and that the variance of each of the
estimated weights is $4\sigma^2/(N + 1)$. It was further indicated [8]
that the estimates of the weights (except that for the bias) are
obtained as orthogonal linear functions of the observations.

In Yates' scheme (1.1), weighing operations 2 through 8
come from the incidence matrix of a symmetrical BIBD characterized by $v = b = 7$, $r = k = 3$, $\lambda = 1$. This is referred to as
the BIBD part of Yates' design. Now, to suit the estimation in
a biased spring balance, if the BIBD part of Yates' design
(which is complementary to L_N, $N = 7$) is augmented by a column of 1's and a row of $(1 + n)$ 1's $(n \leq N)$ and $(N - n)$
0's in that order, $[X'X]$ of the resultant design matrix is
given by

$[X'X]$

$$= \begin{bmatrix} (N+1)I_1 & rE_{1,n} & (r-1)E_{1,N-n} \\ rE_{n,1} & (r-\lambda)I_n + \lambda E_{n,n} & (\lambda-1)E_{n,N-n} \\ (r-1)E_{N-n,1} & (\lambda-1)E_{N-n,n} & (r-\lambda)I_{N-n} + (\lambda-1)E_{N-n,N-n} \end{bmatrix}$$

$$(4.10)$$

where the symbols have the same meanings as in (4.7).

The determinant of (4.10) is given by

$$|X'X| = \left(\frac{N + 1}{4}\right)^{N-1} \left(n - \frac{N - 1}{2}\right)^2 \qquad n \leq N \qquad (4.11)$$

Determinant (4.11) is maximized when $n = N$ and is obtained as
$(N + 1)^{N+1}/4^N$, which is the same as (4.9).

Matrix (4.10), for the special case of n = N, was also
obtained and inverted in the context of providing orthogonal
estimates [8]. The first diagonal element of the inverse is 1
and each of the others is $4/(N + 1)$; these values are the same
as obtained by augmenting an L_N, which is, as mentioned before,
the "most efficient" spring balance design [56]. Here also,
estimates of the weights are obtained as orthogonal linear functions of the observations.

Yates' original example is of the above type and is obtained when N = 7. Thus, Yates' design is D-optimal. In addition, it has the following properties: (a) the variances of
the estimated weights are equal, and (b) the estimates are obtained as mutually orthogonal linear functions of the observations. Comparison with L_N was sought because it is known
[56] that L_N is D-optimal and also provides the minimum variance (equal for each estimated weight). It is thus evident that
no other design better than that of Yates is available which
has all of the above-mentioned desirable properties. The only
other design equivalent to Yates' design is the one obtained by
augmenting the corresponding L_N, as indicated above.

4.5 WEIGHING DESIGNS UNDER AUTOCORRELATION OF ERRORS

4.5.1 Introduction

We have seen in the preceding pages that some weighing
designs have the maximum possible efficiency. These designs
are efficient within the frame of the model when the error
structure is assumed to be of the form $E(\underline{\epsilon}\underline{\epsilon}') = \sigma^2 I_N$. It is
perhaps desirable to examine how the efficiencies of these de-
signs would alter, if at all, if the errors were assumed to be
autocorrelated, that is, if the error structure were assumed to

take the form $E(\underline{\epsilon}\,\underline{\epsilon}') = \sigma^2 V$. (Some results in this direction have been indicated by the author [11].)

Recall that a weighing design is called the best if (a) each variance factor c_{ii} is the least or (b) the average trace of $C = [X'X]^{-1}$, that is, $\Sigma\, c_{ii}/p$, is the least or (c) $\det|A|$ is the maximum or, equivalently, $\det|C|$ is the minimum.

In some cases, criteria (a), (b), and (c) would, of course, lead to an identical situation. However, in the context of reviewing the efficiencies of some of the standard weighing designs under autocorrelation of errors, we shall adopt criterion (b) in defining the efficiency of a weighing design, and we shall denote the efficiency of a design as "trace efficiency."

4.5.2 Structure of Autocorrelation and the Measure of Trace
       Efficiency

The error structure, as mentioned above, is assumed to take the form $E(\underline{\epsilon}\,\underline{\epsilon}') = \sigma^2 V$, where V is given by

$$\begin{bmatrix} 1 & \rho & \rho^2 & \cdots & \rho^{N-1} \\ \rho & 1 & \rho & \cdots & \rho^{N-2} \\ \cdots\cdots\cdots\cdots\cdots\cdots\cdots\cdots\cdots\cdots\cdots \\ \cdots\cdots\cdots\cdots\cdots\cdots\cdots\cdots\cdots\cdots\cdots \\ \rho^{N-1} & \rho^{N-2} & \rho^{N-3} & \cdots & 1 \end{bmatrix}$$

Neither ρ nor its sign is known in practice.

Under the modified model, the least-squares estimates remain the same and are given by $\hat{\underline{\beta}} = CX'\underline{Y}$. $\mathrm{Cov}(\hat{\underline{\beta}})$ is different and is given by $\mathrm{cov}(\hat{\underline{\beta}}) = B = \sigma^2 CX'VXC$.

$\mathrm{Tr}\ B = \sigma^2\ \mathrm{Tr}(MM'V)$

where $M = XC$. In particular, when $N = p$, $\mathrm{Tr}\ B$ is given by

$\mathrm{Tr}\ B = \sigma^2\ \mathrm{Tr}([X'X]^{-1}V)$

4.5.3 Chemical Balance Problem

It was indicated earlier that the best chemical balance design for $p \leq N$ is given by any p columns of the Hadamard matrix H_N, when it exists. In particular, when $p = N$ and the errors are correlated, the trace efficiency of the best chemical balance design is given by σ^2/N.

An individual variance factor may be a function of ρ, but the sum of the variance factors is independent of ρ. In other words, the average trace remains the same even when the errors are autocorrelated. Thus, orthogonal chemical balance designs are the best even when the errors are correlated.

When the errors are not correlated, H_N, as a chemical balance design, is also the best by criterion (c), since $\det|C|$ = $1/N^N$ takes the least possible value. Under autocorrelation, of course, the value of the corresponding determinant is equal to $(1 - \rho^2)^{N-1}/N^N$. When $\rho = 0$, the value of the determinant is equal to $1/N^N$.

4.5.4 Efficiency of a BIBD Under Autocorrelation

It has been indicated [11] that efficiencies of BIBD's, as spring balance weighing designs, depend on the sign of the autocorrelation. The following findings [11] may be of interest:

1. When ρ is positive, BIBD's in general have increased precision compared to the situation when the errors are uncorrelated. An extremely favorable situation is obtained when $\rho \to +1$.

It has been shown with reference to L_N, which represents a special class of BIBD, that when $\rho \to +1$, the average trace reduces to

$$\frac{4\sigma^2}{(N + 1)^2}$$

In particular, in an L_3 each estimate has a variance equal to $\sigma^2/4$, a level of precision obtainable only for the best chemical balance design for $N = 4$. (This is mentioned here because it is known that a spring balance design can be no more than about 1/4 as efficient as a chemical balance design when there is no autocorrelation.)

2. If ρ is negative, the precision is less. An extremely unfavorable situation arises when $\rho \to -1$. Even in this situation, however, a BIBD, as a weighing design, would be preferable to individual weighings.

4.6 SINGULAR WEIGHING DESIGNS

4.6.1 Nature of the Problem

The design matrix X has so far been assumed to be of full rank (rank p). When X is of full rank, matrix $[X'X]$ is nonsingular. But it is possible that, as a result of "bad designing" or despite the best of intentions, the chosen design matrix X will have less than full rank. Consequently, matrix $[X'X]$ will be singular. Thus, one may have to deal with the problem of what can be called a singular weighing design.

It is known that when X is not of full rank, it is not possible to have a unique (unbiased) estimate of each of the objects under nonrandomized procedures (see Zacks [83]). It is possible, however, to have a unique unbiased estimate of an estimable linear function of the objects. Thus in a singular weighing design, it may be necessary to ascertain whether it is possible to furnish an estimate of a given linear function of the objects (say, total weight).

If, after taking the observations in accordance with a design, we detect that the design matrix is singular, we may be

required to take additional weighings to make up for the deficiency in rank. In that case, the problem is to determine how
best we can achieve this augmentation.

4.6.2 Treatment of Singular Weighing Designs

Raghavarao [62] visualized that bad designing, repetitions,
or accidents might lead to singular weighing designs and considered the question of taking additional weighings to make the resultant design matrix X of full rank. He developed a procedure of taking the additional weighings in such a way that the
resultant $\det|X'X|$ would be maximized, as required under
Mood's efficiency definition. For example, the design matrix

$$\begin{bmatrix} 1 & 1 & 1 \\ 1 & 1 & 1 \\ 1 & -1 & 0 \\ 1 & -1 & 0 \end{bmatrix} \tag{4.12}$$

is of rank 2. Therefore, the matrix has to be augmented by one
row to make up for the deficiency. Such a row was obtained by
Raghavarao [62] as (1, 1, -1), which maximizes the value of
$\det|X'X|$ of the resultant design matrix. However, the additional row could also be (-1, -1, 1), since this row also gives
the same maximum value of $\det|X'X|$.

Raghavarao [62] appears to have dealt with the situation
in which the design matrix X is less than full rank by 1. The
procedure can be generalized.

A procedure for making up for the deficiency of rank of
the singular design matrix was also developed by the author [12].
The procedure consisted of augmenting the design matrix by the
inclusion of additional rows under the condition that $\det|X'X|$
of the resultant design matrix be maximized. The methodology

works both in spring balance and chemical balance problems and
in situations where the deficiency in rank is more than 1.

Design (4.12) is such that the third column is half the
sum of the elements in the first two columns. Let us add to it
a fourth column that is half the difference of the elements in
the first two columns. The design matrix will then be of dimensions 4×4 and of rank 2, as shown below:

$$\begin{bmatrix} 1 & 1 & 1 & 0 \\ 1 & 1 & 1 & 0 \\ 1 & -1 & 0 & 1 \\ 1 & -1 & 0 & 1 \end{bmatrix} \qquad (4.13)$$

The two rows to be added to the above design matrix X (4.13)
were obtained [12] as

$$\begin{bmatrix} 1 & 1 & -1 & 1 \\ 1 & -1 & -1 & -1 \end{bmatrix}$$

Total weight is estimable in (4.12). The variance of the
total weight is $\sigma^2/2$. But total weight is not estimable [12]
in design (4.13).

It has been pointed out [12] that, since some linear functions of the objects are estimable in singular weighing designs,
the efficiencies of two singular weighting designs might be compared on the basis of the precision of an estimated linear function of the objects (say, estimated total weight, if estimable),
if the linear function is estimable in both.

It has also been observed that, in spite of very good intentions on the part of an experimenter, a singular weighing
design might be encountered, as the following example shows.

The D_6 obtained by Mood [56] as a spring balance design
for the case $N = p = 6$ by maximizing $\det|X'X|$ was recognized

[10] as a partially balanced incomplete block design. This
PBIBD, when used as a weighing design, is the most efficient
nonsingular spring balance weighing design. This, however, does
not imply that all PBIBD's (when $b \geq v$) give nonsingular weighing designs. This is mentioned because all BIBD's $(b \geq v)$ as
a general rule do furnish efficient nonsingular weighing designs.

   We shall consider, in this context, the following example
of a PBIBD with two associate classes, as developed by Bose and
Nair [25]. The design is given by the nine blocks $(1, 2, 3)$,
$(4, 5, 6)$, $(7, 8, 9)$, $(1, 7, 5)$, $(2, 9, 6)$, $(1, 8, 6)$, $(2, 7, 4)$,
$(3, 9, 5)$, and $(3, 8, 4)$, with the parameters $v = b = 9$, $r =$
$k = 3$, $\lambda_1 = 1$, $n_1 = 6$, $\lambda_2 = 0$, $n_2 = 2$, and

$$p_{ij}^1 = \begin{bmatrix} 3 & 2 \\ 2 & 0 \end{bmatrix} \qquad p_{ij}^2 = \begin{bmatrix} 6 & 0 \\ 0 & 1 \end{bmatrix}$$

The nine varieties (objects) are denoted by the nine integers.
On rearranging the objects $(1, 2, 3, 4, 5, 6, 7, 8, 9)$ in the
order $(1, 4, 9, 2, 8, 5, 3, 7, 6)$, the design matrix takes the
following form:

$$X = \begin{bmatrix}
1 & 0 & 0 & 1 & 0 & 0 & 1 & 0 & 0 \\
0 & 1 & 0 & 0 & 1 & 0 & 0 & 1 & 0 \\
0 & 0 & 1 & 0 & 0 & 1 & 0 & 0 & 1 \\
1 & 0 & 0 & 0 & 1 & 0 & 0 & 0 & 1 \\
0 & 0 & 1 & 1 & 0 & 0 & 0 & 1 & 0 \\
1 & 0 & 0 & 0 & 0 & 1 & 0 & 1 & 0 \\
0 & 1 & 0 & 1 & 0 & 0 & 0 & 0 & 1 \\
0 & 0 & 1 & 0 & 1 & 0 & 1 & 0 & 0 \\
0 & 1 & 0 & 0 & 0 & 1 & 1 & 0 & 0
\end{bmatrix}$$

The columns were rearranged in the above manner to secure some
internal symmetry in the formation of [X'X]. Matrix [X'X]
will have, in the diagonal, three diagonal matrices with 3 in
the diagonal and 0 elsewhere. The other elements of [X'X]
will be 1.

If we add the columns of X three at a time and subtract
the sum of the first three from that of the second and third,
we have two columns of 0's. The design matrix is thus singular.
The rank is 7.

By the procedure developed by the author [12], the two rows
to be added to the design matrix X correspond to the addition
of two blocks, (2, 8, 5) and (3, 7, 6). It is interesting to
observe that if block (1, 4, 9) is also added to the PBIBD,
the design becomes a BIBD with v = 9, b = 12, r = 4, k = 3,
and λ = 1. Of the three blocks added to the PBIBD to make the
design a BIBD, two, namely (2, 8, 5) and (3, 7, 6), are the
same as those obtained through the principle of maximizing the
value of det$|$X'X$|$ of the resultant design matrix. This is
mentioned here because for some BIBD's the value of det$|$X'X$|$,
as we have seen, is maximum.

There was an element of trial and inspection in the procedure of augmentation given by the author [12]. Hazra and
Banerjee [47] have indicated how the augmentation procedure can
be treated on an analytical basis, dependent on the concepts of
the unique Moore-Penrose g-inverse. In the context of singular
weighing designs, the article by Hazra [46] is also of interest.

4.7 FRACTIONAL WEIGHING DESIGNS

4.7.1 Nature of Fractional Weighing Designs

It was mentioned earlier that, for the estimates of
weights of individual objects, the best design for a chemical
balance is given by a Hadamard matrix X of dimensions p × p,

when it exists. If, for lack of resources or time or for other
reasons, only r rows of X are to be used for the weighing
operations, the resultant design matrix is of dimensions r × p
and is a <u>fraction</u> of the full design matrix X. The resultant
matrix is necessarily singular. While it is not possible with
such a (<u>nonrandomized</u>) singular, fractional weighing design to
provide unique and unbiased weights of the individual objects,
it may be possible to find, as is well known, unique unbiased
estimates of some linear functions (estimable) of the objects.
But fractional weighing designs under <u>randomized</u> <u>procedures</u>
provide unbiased estimates for any linear function $\lambda'\underline{\beta}$ (which,
of course, includes any component of $\underline{\beta}$).

4.7.2 Unbiased Estimates Under Randomized Procedures

Zacks [83] visualized the possible use of such randomized,
fractional weighing designs. The fractional weighing design is
obtained by choosing at random n (n < p) rows, independently
and with replacement, from a given Hadamard matrix according to
a probability vector $\underline{\xi}$ of order p. Each of these n rows
specifies a weighing operation to be performed. Given below is
a brief outline of the randomization procedure developed by
Zacks [83] in this context.

"Let $\underline{\xi}$ be a probability vector of order p, and let (Ξ)
be a diagonal matrix of order p, whose diagonal elements are
the components of $\underline{\xi}$. (Ξ) has been called the randomization
procedure matrix. A random choice of one row of (X) is performed according to the probability vector $\underline{\xi}$. Let (J) be a
diagonal matrix representing a choice of one row of (X) whose
diagonal elements are

  $J_i = 1$ if the ith row of X is chosen
    $= 0$ otherwise

(J) has been called an allocation matrix. Obviously, $E\{(J)\}$ $= (\Xi)$. Suppose n rows of (X) are chosen independently, and according to the same probability vector $\underline{\xi}$. Let $(J^{(i)})$ (i = 1, 2, ..., n) denote the allocation matrix of jth weighing operation. Let $\underline{Y}^{(j)}$ (j = 1, 2, ..., n) be identically distributed random vectors, satisfying the model, $\underline{Y} = (X)\underline{\beta} + \epsilon$. The n randomly allocated weighing operations $(1 \leq n \leq p)$ yield the random vectors $(J^{(1)})\underline{Y}^{(1)}, \ldots, (J^{(N)})\underline{Y}^{(n)}$. Let $\underline{Z}^{(n)} = \Sigma_{j=1}^{n} (J^{(j)})\underline{Y}^{(j)}$. The vector $\underline{Z}^{(n)}$ is of order p. The components of $\underline{Z}^{(n)}$ correspond to the randomly chosen components of $\underline{Y}^{(1)}, \ldots, \underline{Y}^{(n)}$. Under the assumption that the distribution function of $\underline{Y}$ is normal, $\underline{Z}^{(n)}$ is a complete sufficient statistic [83]. Denoting by (Ξ^*) a non-singular random procedure matrix corresponding to a probability vector $\underline{\xi}^*$ having positive components, we get

$$\underline{\hat{\beta}}^{(n)}(\underline{\xi}^*) = (np)^{-1}(X)'(\Xi^*)^{-1}\underline{Z}^{(n)}$$

as the unique unbiased linear estimator of $\underline{\beta}$ for every $n \geq 1$."

A randomization procedure has also been developed by Zacks [83] which affords the unbiased estimation for any $\underline{\lambda}'\underline{\beta}$ with minimum variance. The probability vector for the randomization procedure depends, of course, on the linear functional $\underline{\lambda}$. It has, in fact, been indicated by Zacks [83] that every functional $\underline{\lambda}$ specifies a subset of, say, r $(1 \leq r \leq p)$ admissible weighing operations (rows of the Hadamard matrix) in the sense that, if other weighing operations were chosen, the estimation procedure would either be biased or have a variance larger than the one obtainable under admissible weighing operations. If each of the r admissible weighing operations $(1 \leq r \leq p)$ is chosen with probability equal to $1/r$, the corresponding unbiased estimator will have a uniformly (in $\underline{\beta}$ and σ^2) minimum variance.

Finally, the formulas of the unbiased estimator of $\lambda'\beta$ and its variance have been extended to the case of random choice, without replacement, of n rows out of the r admissible rows.

4.7.3 Some Analogous Results Obtained Under Nonrandomized Procedure

Some results of connected interest have been indicated [13] with respect to fractional weighing designs which, of course, are obtained without resorting to any randomization procedure. It has been shown [13] what connection the linear functional λ has with the fractional weighing design and, in that context, it has been pointed out to what extent we could be arbitrary in the selection of the components of λ. The structure of the estimable linear function along with the variance of its estimate has been fully described, bringing out the connection of this variance with the variance obtainable under the full design matrix. It has been shown that with a fraction, which, of course, would depend on λ, we could obtain the same precision for the estimate as would be obtainable in a full design matrix without having to perform all the weighing operations. This fraction is, in fact, the same as given by the set of admissible rows referred to by Zacks [83] under randomization procedure.

4.7.4 Extension of the Procedure to the Spring Balance Problem

Beckman [21] extended the results obtained by Zacks [83] to the spring balance problem. In the development of the procedure, he characterized a design to be the best if the trace of the covariance matrix of the estimator was the minimum. Both randomized procedures and nonrandomized analogs have been studied [21]. The results are similar to those obtained by Zacks [83] and the author [13] in the case of the chemical balance problem.

Later, Beckman [22] was involved in a problem of determining the thermal expansion of plutonium oxide pellets and
visualized the problem to be an application of what he called
"multivariate weighing designs."

4.8 BIASED ESTIMATION IN WEIGHING DESIGNS

4.8.1 Problem of Biased Estimation in Regression

The model for weighing designs is also the model for
multiple regression in general. In regression problems, the
matrix of normal equations is occasionally "ill-conditioned,"
with the consequence that it becomes difficult to solve the
system of normal equations. This difficulty is, in some situations, circumvented by the addition of a small positive quantity k to the diagonal elements of matrix $[X'X]$. When such
a small quantity is added, the normal equations take the form
$[X'X + kI]\underline{\hat{\beta}}^* = X'\underline{Y}$. The estimator $\underline{\hat{\beta}}^*$ obtained from solving
these equations is obviously biased.

Hoerl and Kennard [48] have studied a procedure of such
biased estimation in the regression problem in general and have
indicated how the value of k can be chosen so that the total
mean square error of $\underline{\hat{\beta}}^*$ is less than the total variance of
the least-squares unbiased estimator $\underline{\hat{\beta}}$.

It has also been shown [48] that the length of the biased
estimator $\underline{\hat{\beta}}^*$ is shorter than the length of the unbiased estimator $\underline{\hat{\beta}}$.

4.8.2 Results Obtained in Weighing Designs

Sihota and Banerjee [67] studied the effect of such a
biased estimation procedure in weighing designs, in which the
design matrices are of a special nature, the elements x_{ij} of

X being either $(\pm 1, 0)$ in the chemical balance or $(1,0)$ in the spring balance. They also compared biased estimates with the unbiased estimates, and the mean square error with the variance [67].

Some of the results indicated by Sihota and Banerjee [67] with reference to weighing designs also have a bearing on the general problem of biased estimation in regression analysis [48]. We indicate below, as an illustration, only one of the results reported by Sihota and Banerjee [67].

The coefficient of variation of a biased estimator $\hat{\beta}_i^*$ derived from an orthogonal chemical balance design matrix X (which, of course, corresponds to the "orthogonal" problem in general regression) has been found to be larger than the coefficient of variation of an unbiased estimator $\hat{\beta}_i$, although the total mean square error of $\underline{\hat{\beta}}^*$ is less than the total variance of $\underline{\hat{\beta}}$. This happens because the decrease in the mean square error of $\underline{\hat{\beta}}^*$ does not offset the decrease in the length of $\underline{\hat{\beta}}^*$. However, if, in the calculation of the coefficient of variation of the biased estimator, the square root of the variance is used in place of the root mean square error, the coefficient of variation of a biased estimator $\hat{\beta}_i^*$ remains the same as that of an unbiased estimator $\hat{\beta}_i$.

4.9   REPEATED SPRING BALANCE WEIGHING DESIGNS

When a symmetrical BIBD is used as a spring balance weighing design to estimate the weights of $N = p$ objects, no degrees of freedom are left for the estimation of the error variance. In such a situation, the symmetrical BIBD could perhaps be repeated [1] in order to secure "degrees of freedom" for the estimation of the error variance. However, Dey [31] has pointed

out that, if in a symmetrical BIBD we have b > 2r, then we
should combine the symmetrical BIBD with its complementary BIBD
rather than repeat it. Again, to meet the same requirement,
Kulshreshtha and Dey [54] have suggested an alternative design
which is preferable to "repeated designs" and to those suggested
by Dey [31], provided one is interested in estimating the weights
of some of the objects with increased precision at the cost of
precision for others. However, it has been demonstrated [16]
that, with a given problem, one may repeat the complementary
BIBD.

4.10 THE CONSTRUCTION OF OPTIMAL CHEMICAL BALANCE DESIGNS

Given the restraints that the individual variances be
equal and that the covariances between pairs also be equal,
that is, subject to $[X'X]$ being of the form $M(r, \lambda)$,
Raghavarao [60, 61] showed that, when N is odd, $[X'X]$ of a
D-optimal design has to be of the form $M(N, 1)$ and that, when
$N \equiv 2 \pmod 4$, $[X'X]$ of an E-optimal design has to be of the
form $M(N - 1, 0)$. The first series of these designs is denoted
by P_N, and the second series by S_N. Raghavarao [60, 61] constructed the P_N series from the incidence matrix of a BIBD of
a special kind by changing the zeros to -1. The S_N series
was constructed following a method of Williamson [71-73]. Again,
when $N \equiv 2 \pmod 4$ and $N \neq 2$, the D-optimal design is of the
form for which $[X'X] = M(N, 2)$. Such designs exist [61] only
for $N = 6$ and $N = 66$ out of all $N < 200$.

Given the same restraints, Bhaskararao [23] constructed
several additional series of D-optimal chemical balance designs
when N is odd, along with optimal designs when $N \equiv 2 \pmod 4$.
Bhaskararao [23] denoted such designs by $[N, s, \lambda]$, characterized by the parameters N, s, and λ, where N is the size

(order) of the square design matrix X; s is the number of
zeros in any column of X; and $\lambda = \Sigma_{i=1}^{N} x_{ij} x_{ij'}$, [j, j' (j $\neq$ j')
= 1, 2, ..., N]. Legendre symbols, Hilbert norm residues, and
the Hasse-Minkowski invariants have been used, as usual, to
prove the nonexistence of designs in these categories for some
values of the parameters N, s, and λ.

Dey [32] has also presented methods of constructing optimal chemical balance designs from the incidence matrix of a BIBD
by substituting -1 for 0. In particular, he has shown that
chemical balance designs with the minimum possible variance
can be obtained from the incidence matrix of a BIBD when v $\neq$ 2k
and b = 4(r - λ). For such designs, v = $[(4k+1)\pm(8k+1)^{1/2}]/2$.
When v = 2k, the weighing design becomes singular. Dey [32]
has also shown how such singular designs can be adjusted to obtain designs with the minimum possible variance.

4.10.1 A Comparison of E-optimal and D-optimal Designs

We shall now show how a D-optimal design (with the restraints as stated above) differs from a D-optimal design without any restriant and an E-optimal design. Given below are examples of an E-optimal design and a D-optimal design with the
restraints stated above:

E-optimal: D-optimal (under restraints):

N = p = 6   N = p = 6
(An illustration given by   (An illustration given by
 Raghavarao [60] and   Raghavararo [61])
 Banerjee [7])

	I					

$$
\begin{bmatrix}
0 & 1 & 1 & 1 & 1 & 1 \\
1 & 0 & 1 & -1 & -1 & 1 \\
1 & 1 & 0 & 1 & -1 & -1 \\
1 & -1 & 1 & 0 & 1 & -1 \\
1 & -1 & -1 & 1 & 0 & 1 \\
1 & 1 & -1 & -1 & 1 & 0
\end{bmatrix}
$$

II

$$
\begin{bmatrix}
1 & -1 & -1 & -1 & -1 & -1 \\
1 & -1 & 1 & 1 & 1 & 1 \\
1 & 1 & -1 & 1 & 1 & 1 \\
1 & 1 & 1 & -1 & 1 & 1 \\
1 & 1 & 1 & 1 & -1 & 1 \\
1 & 1 & 1 & 1 & 1 & -1
\end{bmatrix}
$$

(i) $|X'X| = 5^6$.   (i) $|X'X| = 2^{14}$.

(ii) The estimates are   (ii) The estimates are
      equally correlated.   equally correlated.

(iii) $c_{ii} = 1/5$.   (iii) $c_{ii} = 7/32$.

The following is an example of a D-optimal design (without any restraints):

D-optimal (unrestricted):
$N = p = 6$
(An illustration given by Mood [56])

III

$$\begin{bmatrix} 1 & -1 & -1 & -1 & -1 & -1 \\ 1 & -1 & -1 & -1 & 1 & 1 \\ 1 & -1 & -1 & 1 & 1 & -1 \\ 1 & -1 & 1 & 1 & -1 & 1 \\ 1 & 1 & 1 & -1 & 1 & -1 \\ 1 & 1 & -1 & 1 & -1 & 1 \end{bmatrix}$$

(i) $|X'X| = 5^2 2^{10}$.

(ii) The estimates are not
       equally correlated.

(iii) $c_{ii} = 1/5$.

Notice that the variance factors in design III are smaller than those in design II and that, although $|X'X|$ for III is larger than that for I, the variance factors are the same.

From all considerations, therefore, design III is preferred, since it has the maximum possible value of $|X'X|$ and its variance factors are no larger than those of either of the other two designs.

4.11 CONSTRUCTION OF D-OPTIMAL (UNRESTRICTED) DESIGNS

Ehlich [34], Ehlich and Zeller [36], and Yang [74-77] have constructed square matrices of orders $N \equiv 1, 2 \pmod 4$ such that the values of the corresponding determinants are

maximum. These square matrices are D-optimal without any restraints. Design III (Sec. 4.10.1) is a square matrix of this
type. That this square matrix (design III) has the maximum
possible value of the determinant is evident from the following
inequality. Ehlich [34] has shown that the maximal absolute
value α_N of the Nth-order determinant with entries ± 1 satisfies

$$\alpha_N^2 \leq 4(N - 2)^{N-2} (N - 1)^2$$

When $N = 6$, the maximum possible value of α_N^2 is $5^2 2^{10}$, as
attained by design III.

4.12 A NEW MODEL FOR THE WEIGHING PROBLEM AND A COMMENT ON THE CONSTRUCTION OF BALANCES

The estimation of the weights is based on the assumption
that the weight pieces that come with the balance represent the
true weights for which the weight pieces are marked. True weights
are never known. Depending on the manufacturing process, minor
variations in the marked weights are likely. With this question
in mind, a new model for the weighing problem has been suggested
[19] to indicate how such weight piece errors may affect the precision of the estimated weights.

Let $\underline{W} = (W_1, W_2, \ldots, W_q)'$ be the marked weights of the
weight pieces used in any weighing experiment, and let $\underline{f} =
(f_1, f_2, \ldots, f_q)'$ be the associated random errors. We assume
$E(\underline{f}) = \underline{0}$, $E(\underline{ff}') = \text{diag}(\sigma_1^2, \ldots, \sigma_q^2)$, and $E(e_i f_k) = 0$ (i =
1, 2, \ldots, N; k = 1, 2, \ldots, q). In observing a y, there is
an experimental error. Each y, again, is an algebraic sum of
the marked weights of the weight pieces. Thus, $\underline{Y}$ is taken as
equal to ZW, where $Z = (z_{ik})$ is an $N \times q$ matrix of elements
± 1, 0, while $\underline{Y}$ is actually equal to $Z(W + f)$. Hence, the
model is

$$\underline{Y} = X\underline{\beta} + Z\underline{f} + \underline{e}$$

In the setup of the above model, a few theorems were proven by the author [19], including one to show that suitable designing cannot reduce the total weight piece error to zero.

After a period of time, while natural wear and tear may cause some fluctuation to the weights of the weight pieces, accumulation of dirt may also cause some fluctuation in the opposite direction. If these two effects cancel each other out, the weight pieces will have the same status as when manufactured. If, on the contrary, there is a resultant effect in one direction or the other, the estimates will be biased. The question of such bias is dealt with [19] by incorporating an additional component in the model.

Let, after a period of time, the effect of wear and tear in one direction, and the accumulation of dirt in the other, bring about a resultant effect of $\underline{\mu} = (\mu_1, \ldots, \mu_q)'$ in the weight pieces, where μ_k $(k = 1, 2, \ldots, q)$ is an unknown constant. The model will then be modified:

$$\underline{Y} = X\underline{\beta} + Z\underline{\mu} + Z\underline{f} + \underline{e}$$

For this modified model, the least-squares estimates $\hat{\beta}^*$ and the mean square error (MSE) for $\hat{\beta}^*$ are shown by the author [19]. How large the bias and the MSE may be are also indicated. Eventually, a tentative comment is made on the manufacturing practices of the balances, suggesting combinations of weight pieces within certain ranges. (The problem under consideration indeed calls for a more detailed investigation than attempted by the author [19].)

4.13 USE OF WEIGHING DESIGNS IN THE MEASUREMENT OF
PHYSICAL CONSTANTS

In a paper addressed to both physicists and statisticians,
Youden [81] discussed the use of various experimental designs to
increase the precision of measurement of physical constants
through the detection of systematic errors.

In the measurement of a physical constant, an experimenter
may be involved in evaluating the differential effects of a number of adjustments and substitutions of components of the apparatus. If he tried to evaluate these effects "one by one," as
one might ordinarily do, the required number of observations
would be too large, because in the interest of precision, a con-
siderable number of repeat measurements must be made for each
assembly of the components of the apparatus and for each of the
adjustments considered.

Let us suppose, for example, that an experimenter has seven
alternatives (including both adjustments and substitutions for
components) under consideration. Under the "one-by-one" technique the experimenter "may designate some standard initial assembly and set of adjustments and then proceed to change, one by
one, the seven items under consideration. Some measurements are
made under the initial state, an item is changed and another set
of measurements made. Whatever was changed is put back to the
initial state and a second item changed. There will be eight
such sets and a goodly number of measurements are required in
each set."

However, Youden observed that, through the principles of
designing as developed under weighing designs, the way is open
to change more than one variable at a time, provided that the
effect of a small change in a variable does not depend on the
other variables or, in other words, provided that there is no
interaction among the variables (to speak in the language of
factorial experiments).

If we admit that there are two possible variations (or
states) for each of the seven items t, u, v, w, x, y, and z,
the design suggested by Youden for studying simultaneously the
effects of seven items is given by the following scheme:

1	2	3	4	5	6	7	8
t	t	t	t	T	T	T	T
u	u	U	U	u	u	U	U
v	V	v	V	v	V	v	V
w	w	W	W	W	W	w	w
x	X	x	X	X	x	X	x
y	Y	Y	y	y	Y	Y	y
z	Z	Z	z	Z	z	z	Z

In this setup, the capital letters stand for the second variation (or state) of the items.

This design is easily recognized as the weighing design
given by Yates [78], the design that originated the weighing
problem. To recognize that these two designs are the same, the
capital letters T, U, V, W, X, Y, and Z (second states) can
be substituted by blanks (denoting absence of the items, or
"control"). If these capital letters are substituted by t_0,
u_0, v_0, w_0, x_0, y_0, and z_0, respectively (denoting absence,
or "control"), to accord with the notation of factorial experiments, the above design is recognized as a fractional replicate
of 1 in 16 of a 2^7-factorial experiment (see Sec. 4.2).

Estimates of the effects of each of the seven items are
obtained the same way the weights of seven light objects are
obtained. For example, the effect $(t - T)$ is given by
$[(1 + 2 + 3 + 4) - (5 + 6 + 7 + 8)]/4$, which, in fact, is the
estimated weight of the object t. These seven estimates have
all the properties referred to earlier, including precision
obtained through combination of operations. Youden's illustration thus serves as a reminder of Hotelling's observation (Sec.
1.2.2).

4.14. USE OF WEIGHING DESIGNS IN OTHER SCIENCES

Experimental designs in various forms, depending on the problem, are used in nearly all applied sciences. How optimal designing can be of help in general has been amply demonstrated by the work of Box and his associate [26-28].

Reference can also be made to the work of Broyles and Eckert [29] on planned experiments in studies on sulfonation of toluene and to the work of Sane et al. [66] on the sequential design of experiments for chemical kinetic modeling to indicate how the design of experiments can come to our aid in chemistry and chemical engineering.

It is well known how useful in industrial research are the optimum main effect plans (i.e., plans to determine the main effects with the minimum number of experiments) in 2^n factorials. Such plans, as indicated by Plackett and Burman [59], are provided by Hadamard matrices which, as we have seen, give the most efficient chemical balance designs. Hadamard matrices and the associated (0, 1) matrices thus provide the link between the special class of designs known as weighing designs, on the one hand, and designs in general, on the other.

In Sec. 4.13, we saw how a weighing design was utilitized by Youden [82] in the determination of physical constants.

Recently, Sloane, et al. [68, 69] and Harwit [45] (and possibly many others), have started using Hadamard matrices and the associated (0, 1) matrices in their investigations in multiplex spectrometry. The authors have not called these designs weighing designs,* although in fact they are.

---

*In a letter dated 28 August, 1974, addressed to the author, Dr. Sloane has written as follows: "You might be interested in the enclosed, which gives an application of weighing designs (only we didn't call them by that name!) in measuring frequency spectra."

The references in this section indicate that weighing designs have the potential of being useful in the planning of experiments in other sciences, and not merely in determining the
weights of light objects.

4.15 CONCLUDING REMARKS: THE FUTURE OF THE
WEIGHING PROBLEM

It is evident from a perusal of the preceding pages that
there is a need for chemical and spring balance designs of square
and rectangular matrices of higher dimensions such that $\det|X'X|$
is maximized or trace of $[X'X]^{-1}$ is minimized without any restriction on the equality (or otherwise) of the variances and
covariances. It is hoped that some progress will be made in
this direction in the future.

The scope of the problem will perhaps be broadened when
other side issues are resolved to meet ad hoc needs. For example, one might like to know the form the error structure would
take if the "objects" were not necessarily light and, knowing
the error structure, one might like to know how one should proceed in the selection of the best design.

On the other hand, one might legitimately ask to what extent these weighing designs have been found, or will be found,
useful in practice. Professor Hotelling once told the author
that he knows of some chemists who have been utilizing these
designs in their day-to-day routine. Possibly, many others in
other sciences, including operations research analysts, are
also utilizing weighing designs in this way. Perhaps, some are
making use of the principles involved. The author himself once
used these principles in an estimation problem in a socioeco-
nomic survey. He feels hopeful, however, that these designs
will eventually attract the wider attention of research workers

in all applied sciences and that these designs will, in the
future, find their rightful place in estimation problems and
in laboratory work where weighing operations are needed as a
matter of routine.

In conclusion, the author would like to quote the following lines from W. J. Youden [81, p. 118].* "I confess, with some
embarrassment, that for at least three investigations over the
past years at the NBS (National Bureau of Standards, U.S.A.),
I suggested individual and special programs and overlooked the
general character and wide applicability of these weighing designs." In this context, the footnote in Sec. 4.14 may also be
of interest.

---

*This article originally appeared in <u>Physics</u> <u>Today</u>,
September, 1961.

1. Set up a statistical model to determine the weight of one
light object by repeated weighings. Given that N weighing operations have been made for the purpose, show that
the least-squares estimate of the weight is given by the
average $\bar{y} = \Sigma_{i=1}^{N} y_i/N$, where y_i $(i = 1, 2, \ldots, N)$ is
the record of the ith weighing. Assuming that each y is
observed with an error ϵ with variance σ^2 and that the
errors are uncorrelated, calculate the variance of this estimate.

2. Determine the normal equations (estimating equations) for
providing the least-squares estimates of β_0 and β_1 of
the linear functional relationship given by $y = \beta_0 + \beta_1 x$.
Assuming that N observations (y_i, x_i) $(i = 1, 2, \ldots, N)$
are taken for the purpose, where each y is observed with
an error ϵ with variance σ^2 and x is fixed, obtain
the estimates of β_0 and β_1 and the variance of the estimates obtained. (Assume that x can take any real value.)

3. Obtain the least-squares estimates of β_0 and β_1 and the
variance of each of the estimates when only two observations
$(y_1, +1)$ and $(y_2, -1)$ have been taken under the model referred to in Exercise 2.

4. State the conditions under which the system of linear equa-
    tions $A\underline{x} = \underline{l}$ will have solutions for $\underline{x}$. Here, A is an
    $n \times n$ matrix of known coefficients, $\underline{x}$ is an $n \times 1$ vector
    of the variables, and $\underline{l}$ is an $n \times 1$ vector of known con-
    stants.

5. Assuming that the weighing problem is characterized by the
    general linear hypothesis model, show that the variance of
    each estimated weight with N weighing operations is mini-
    mum if and only if $[X'X] = NI_p$, where X is the design
    matrix, I_p is the identity matrix of order p, and p is
    the number of objects, the weights of which are to be deter-
    mined. (Hotelling [49] and Moriguti [57].)

6. In a weighing experiment to determine the weights of four
    objects, the design matrix X and the vector of observa-
    tions $\underline{Y}$ were obtained as shown below:

$$X = \begin{bmatrix} +1 & +1 & +1 & +1 \\ -1 & +1 & -1 & +1 \\ -1 & -1 & +1 & +1 \\ +1 & -1 & -1 & +1 \end{bmatrix} \quad \underline{Y} = \begin{bmatrix} \text{grams} \\ 7.370 \\ -0.789 \\ -2.584 \\ +0.041 \end{bmatrix}$$

    (a) Obtain the estimates of the weights.
    (b) Calculate the variance of each estimated weight.
    (c) If each of the four objects were weighed separately
        as in the model of Exercise 1, how many weighing oper-
        ations would be needed in all to secure the same vari-
        ance as obtained under the above design?
    (d) Is it possible to provide an estimate of the variance
        σ^2 from this design? If not, how would you alter or
        augment the design matrix?

7. Let $M(r, \lambda)$ be a $p \times p$ matrix with r in the diagonal and
    λ elsewhere $(r > \lambda)$. Determine the value of the determi-
    nant of $M(r, \lambda)$ and the inverse of the matrix $M(r, \lambda)$.

8. In a spring balance (one-pan balance) weighing design, the
    design matrix X and the vector of observations (records
    of weighings) $\underline{Y}$ were obtained as shown below:

$$
X = \begin{bmatrix} 1 & 1 & 0 \\ 1 & 0 & 1 \\ 0 & 1 & 1 \end{bmatrix} \quad \underline{Y} = \begin{bmatrix} 4.50 \\ 3.50 \\ 5.52 \end{bmatrix}
$$

grams

    (a) Obtain the normal equations for estimating the weights
         of the three objects.
    (b) Find the estimates of the weights.
    (c) Calculate the variance for each of the estimated
         weights.
    (d) Calculate the covariance between a pair of estimates.

9. If, in a weighing design, the design matrix X is such
    that $[X'X]$ is of the form $M(r, \lambda)$ $(r > \lambda)$, as referred
    to in Exercise 7, what would be the variance of each esti-
    mated weight and the covariance between a pair of the esti-
    mates?

10. Taking the efficiency of a weighing design as

$$
\frac{\sigma^2/N}{\sigma^2 \, \Sigma_{i=1}^{p} c_{ii}/p} = \frac{p}{N} \sum_{i=1}^{p} c_{ii}
$$

    where c_{ii} is the ith diagonal element of $C = [X'X]^{-1}$,
    calculate the efficiency of each of the following designs
    (Chakrabarti [30, p. 115]):

$$
(a) \; X = \begin{bmatrix} 1 & 1 \\ 1 & 1 \\ 1 & 1 \\ 1 & -1 \\ 1 & -1 \end{bmatrix} \quad (b) \; X = \begin{bmatrix} 1 & 1 \\ 0 & 1 \\ 1 & 1 \\ 1 & -1 \\ -1 & -1 \end{bmatrix} \quad (c) \; X = \begin{bmatrix} 1 & 1 & 1 & 0 \\ 1 & 1 & 0 & 1 \\ 1 & 0 & 1 & 1 \\ 0 & 1 & 1 & 1 \end{bmatrix}
$$

11. The efficiency of a weighing design X may also be defined
     as $\det|X'X|/\max \det|X'X|$. On the basis of this criterion,
     compare the efficiencies of designs (a) and (b) in Exercise
     10.

12. Taking the criterion of efficiency of a weighing design as
     provided in Exercise 10, find the efficiency of the weigh-
     ing design

$$X = \begin{bmatrix} E_{1p} \\ \\ H_{N-1} \end{bmatrix}$$

where E_{1p} is a $1 \times p$ matrix of unities and H_{N-1} is a
matrix of $N - 1$ rows and p columns of +1 and -1, such
that $H'_{N-1}H_{N-1} = (N - 1)I_p$ (Chakrabarti [30, p. 115]).

13. (a) Show that the efficiency of a weighing design given
     by a BIBD (balanced incomplete block design characterized
     by the parameters v, b, r, k, λ) is $k^2(p-k)/p(pk-2k+1)$,
     where the criterion of efficiency is the same as provided
     in Exercise 10. (Note that p takes the place of v in
     weighing designs.) (Banerjee [1].)
     (b) On the basis of the same criterion of efficiency, find
     the efficiency of a weighing design in which all possible
     $\binom{p}{k}$ weighing operations are performed to find the weights
     of p objects in a spring balance (one-pan balance).
     (Banerjee [1].)

14. If the criterion of efficiency is taken as defined in Exer-
     cise 11, how would you compare the efficiencies of the two
     designs referred to in Exercises 13a and 13b?

15. Show that the variance of the estimated total weight obtained from a BIBD used as a weighing design is given by $p\sigma^2/[r + (p - 1)\lambda] = p\sigma^2/rk$. (Recall that v is the same as p, denoting the number of objects.) (Banerjee [1].)

16. Construct an L_7 from the 8×8 Hadamard matrix H_8 listed in Appendix C (Mood [56]).

17. Use an L_7 (Appendix C) as a weighing design and provide the estimates of weights of the seven objects as linear functions of the observations $y_1, y_2, \ldots, y_7$. Indicate how such estimates can be obtained from the $(1, 0)$ distribution of such design matrices. (Banerjee [8].)

18. Let an L_7 be bordered by a column of $+1$'s and a row of 0's, in that order, and let this additional column occupy the first position of the augmented design matrix. Use this augmented design matrix to provide the estimates of weights of the objects in terms of $y_0, y_1, \ldots, y_7$, where y_0 corresponds to the first row of the augmented design matrix. How do these estimates differ from the estimates in Exercise 17? (Banerjee [8].)

19. An L_7 (listed in Appendix C) was used to determine the weights of the following seven coins:[*] (a) U.S. 25¢, (b) U.S. 5¢, (c) U.S. 1¢, (d) Swedish 5 öre, (e) U.S. 10¢, (f) Swedish 10 öre, and (g) Japanese 1 yen. The seven records of weighings (in grams) were 11.971, 10.399, 7.371, 14.220, 12.983, 12.760, and 14.262, corresponding, respectively, to the rows from top to bottom. Estimate the weights of the coins.

---

[*]These weighing operations were performed by Mrs. Takako Nagase.

20. The BIBD given by $r = 7$, $v = 8$, $k = 4$, $b = 14$, $\lambda = 3$
    (Appendix C) was used as a weighing design to determine
    the weights of eight light objects, and the 14 records of
    weighings (in grams), in order of the rows from top to
    bottom, were 13.11935, 12.49018, 10.32219, 10.32896,
    5.79679, 6.42530, 8.59399, 8.58759, 12.07839, 10.76119,
    11.07992, 6.83760, 8.15368, and 7.83867. Obtain the esti-
    mates of weights of the objects and provide an estimate of
    the variance σ^2 from the data.

21. Find the inverse of the augmented design matrix referred
    to in Exercise 18. (Banerjee [8].)

22. A chemist used an L_7 to determine the weights of seven
    light objects in a spring balance, and later it was brought
    to his attention that the balance he had used was biased
    and needed a zero correction. Can he still make use of the
    records of weighings he made before with the minimum of
    additional effort? If so, how should he proceed, and what
    would be the nature of the revised estimates?

23. An L_7 is given by the arrangement of a special type of
    symmetrical BIBD given by $r = k = (N + 1)/2$, $v = b = N$,
    $\lambda = (N + 1)/4$. If an L_N is used as a weighing design,
    it is known that the variance factor of each of the esti-
    mated weights is the minimum, being equal to $4N/(N + 1)^2$.
    Show that there is no other BIBD for which the variance
    factors would reach this minimum. (Banerjee [3].)

24. Calculate the efficiencies of the two weighing designs
    given below on the basis of the criteria referred to in
    Exercises 10 and 11. What comments do you have? (Banerjee
    [7].)

(a) $X = \begin{bmatrix} 1 & 1 & 1 & 1 & 1 & 1 \\ 1 & 1 & 1 & -1 & -1 & -1 \\ 1 & 1 & -1 & -1 & 1 & 1 \\ 1 & -1 & -1 & 1 & -1 & 1 \\ 1 & -1 & 1 & -1 & 1 & 1 \\ 1 & -1 & -1 & 1 & 1 & -1 \end{bmatrix}$

(b) $X = \begin{bmatrix} 0 & 1 & 1 & 1 & 1 & 1 \\ 1 & 0 & 1 & -1 & 1 & -1 \\ 1 & 1 & 0 & 1 & -1 & -1 \\ 1 & -1 & 1 & 0 & -1 & 1 \\ 1 & 1 & -1 & -1 & 0 & 1 \\ 1 & -1 & -1 & 1 & 1 & 0 \end{bmatrix}$

25. Let the error structure in a weighing experiment be given
as $E(\underline{\epsilon}\underline{\epsilon}') = \sigma^2 V$, where V is known as

$$V = \begin{bmatrix} 1 & \rho & \rho^2 & \cdots & \rho^{N-1} \\ \rho & 1 & \rho & \cdots & \rho^{N-2} \\ \cdots\cdots\cdots\cdots\cdots\cdots\cdots\cdots\cdots\cdots \\ \rho^{N-1} & \rho^{N-2} & \rho^{N-3} & \cdots & 1 \end{bmatrix}$$

With this autocorrelated model:

(a) Obtain the least-squares estimator of $\underline{\beta}$ (call it
$\hat{\underline{\beta}}_L$).

(b) Find the generalized least-squares estimator for $\underline{\beta}$
(call it $\hat{\underline{\beta}}_g$).

(c) Find $\text{cov}(\hat{\underline{\beta}}_L)$.

(d) Find $\text{cov}(\hat{\underline{\beta}}_g)$ (Appendix A).

(e) What happens to $\hat{\underline{\beta}}_L$ and $\hat{\underline{\beta}}_g$ and to $\text{cov}(\hat{\underline{\beta}}_L)$ and
$\text{cov}(\hat{\underline{\beta}}_g)$ when the design matrix X is square and
nonsingular?

(f) Show that, with reference to a Hadamard matrix used
as a weighing design, the least-squares estimator $\hat{\underline{\beta}}_L$ is
still the best in the sense that the average of the vari-
ance factors will be the minimum, it remaining the same
with this model.

26. Given below are two singular weighing designs:

$$
X = \begin{bmatrix} 1 & 1 & 1 \\ 1 & 1 & 1 \\ 1 & -1 & 0 \\ 1 & -1 & 0 \end{bmatrix} \quad X = \begin{bmatrix} 1 & 1 & 1 \\ 1 & -1 & 0 \\ 1 & -1 & 0 \\ 1 & -1 & 0 \end{bmatrix}
$$

(a) Indicate if total weight is estimable in both designs
and, if so, find for both the estimate of the total
weight as well as the variance of the estimated total
weight.

(b) If the purpose were to estimate the total weight,
which of the two designs would you prefer?

(c) Find one additional row for the design matrix (1.1)
to make up for the deficiency in rank of X such
that $\det|Z'Z|$ is maximized, where Z is the aug-
mented design matrix. (Raghavarao [62] and Banerjee
[12].)

Appendix A

THE GENERAL LINEAR HYPOTHESIS MODEL
AND THE METHOD OF LEAST SQUARES

A.1 DESCRIPTION OF A LINEAR MODEL

Let us suppose that a set of mathematical variables x_1, x_2, ..., x_p and the variable y, which is observed with an unknown random error ϵ, are related as

$$y = \beta_1 x_1 + \beta_2 x_2 + \cdots + \beta_p x_p + \epsilon \qquad (A.1)$$

where β_j $(j = 1, 2, ..., p)$ are unknown parameters (constants). Equation (A.1), along with the assumptions about ϵ stated in Sec. A.2, represents a linear model. [Incidentally, by a linear model we mean a mathematical equation of the above type, an equation that involves random variables $(y$ and $\epsilon)$, mathematical variables $(x_1, x_2, ... , x_p)$, and parameters $(\beta_1, \beta_2, ..., \beta_p)$ with linearity in the random variables and the parameters.]

A.2 MATRIX REPRESENTATION OF THE MODEL

In an actual experiment, when all the required observations are taken in accordance with the model, the observational equations are expressed in matrix notation as

$$\underline{Y} = x\underline{\beta} + \underline{\epsilon} \tag{A.2}$$

where

$$\underline{Y} = \begin{bmatrix} y_1 \\ y_2 \\ \vdots \\ y_N \end{bmatrix} \quad X = (x_{ij}) = \begin{bmatrix} x_{11} & x_{12} & \cdots & x_{1p} \\ x_{21} & x_{22} & \cdots & x_{2p} \\ \multicolumn{4}{c}{\cdots\cdots\cdots\cdots\cdots} \\ x_{N1} & x_{N2} & \cdots & x_{Np} \end{bmatrix}$$

$$\underline{\beta} = \begin{bmatrix} \beta_1 \\ \beta_2 \\ \vdots \\ \beta_p \end{bmatrix} \quad \underline{\epsilon} = \begin{bmatrix} \epsilon_1 \\ \epsilon_2 \\ \vdots \\ \epsilon_N \end{bmatrix}$$

In the above setup with N observations, $\underline{Y}$ is an $N \times 1$ column vector of observed random variables y_i ($i = 1, 2, \ldots, N$); $x = (x_{ij})$ ($i = 1, 2, \ldots, N$; $j = 1, 2, \ldots, p$); p ($\leq N$) is a known $N \times p$ matrix with x_{ij} as its (i, j)th element; $\underline{\beta}$ is a $p \times 1$ vector of unknown parameters β_j ($j = 1, 2, \ldots, p$); and $\underline{\epsilon}$ is an $N \times 1$ vector of unobserved random errors ϵ_i ($i = 1, 2, \ldots, N$).

We shall assume for present purposes that the distribution of the random vector $\underline{\epsilon}$ is such that $E(\underline{\epsilon}) = \underline{0}$, and $\text{cov}(\underline{\epsilon}) = E(\underline{\epsilon}\underline{\epsilon}') = \sigma^2 I_N$, where E stands for mathematical expectation, and $\text{cov}(\underline{\epsilon})$ stands for the variance-covariance matrix of the random vector $\underline{\epsilon}$; σ^2 is the variance of ϵ_i $(i = 1, 2, \ldots, N)$. From the form of $\text{cov}(\underline{\epsilon})$, it is evident that the errors ϵ_i $(i = 1, 2, \ldots, N)$ are uncorrelated.

Matrix X is referred to as the <u>design matrix</u>. When the rank of matrix X is full, that is, where rank of X = p $(p \leq N)$, the above model is referred to as the general linear hypothesis model of full rank.

The above setup paves the background for the celebrated Gauss-Markoff theorem on least-squares estimates for β_j $(j = 1, 2, \ldots, p)$.

A.3 LEAST-SQUARES ESTIMATES

The sum of squares (s.s.) of the errors is given by

$$\underline{\epsilon}'\underline{\epsilon} = (\underline{Y} - X\underline{\beta})' \, (\underline{Y} - X\underline{\beta}) = \underline{Y}'\underline{Y} - 2\underline{\beta}'X'\underline{Y} + \underline{\beta}'X'X\underline{\beta} \qquad (A.2)$$

where a prime denotes a transpose. Least-squares estimates of β_j $(j = 1, 2, \ldots, p)$ are those that make (A.2) the least. The value of $\underline{\beta}$ that minimizes (A.2) is given by the solution of the equations

$$\frac{\partial}{\partial \underline{\beta}} \left(\underline{\epsilon}'\underline{\epsilon} \right) = 0 \qquad\qquad\qquad (A.3)$$

where $\partial/\partial\underline{\beta}$ stands symbolically for the column vector of derivatives given by

$$\frac{\partial}{\partial\underline{\beta}} = \begin{bmatrix} \dfrac{\partial}{\partial\beta_1} \\[2ex] \dfrac{\partial}{\partial B_2} \\[1ex] \vdots \\[1ex] \dfrac{\partial}{\partial\beta_p} \end{bmatrix}$$

We now quote two results on symbolic differentiation of quadratic forms and bilinear forms. If $A = (a_{ij})$ is an $n \times n$ symmetric matrix whose (i, j)th element is a_{ij}, and $\underline{Y}$ is an $n \times 1$ column vector with y_i $(i = 1, 2, \ldots, n)$ as its ith element, then $\underline{Y}'A\underline{Y}$ is defined as a quadratic form equal to $\Sigma_{i=1}^{n} \Sigma_{j=1}^{n} y_i y_j a_{ij}$. If $\underline{L}$ is an $n \times 1$ column vector with ℓ_i $(i = 1, 2, \ldots, n)$ as its ith element, then $\underline{Y}'\underline{L}$ is defined as a bilinear form, $\Sigma_{i=1}^{n} y_i \ell_i$. It can be verified that

(i) $\frac{\partial}{\partial \underline{Y}} (\underline{Y}'A\underline{Y}) = 2A\underline{Y}$

(ii) $\frac{\partial}{\partial \underline{Y}} (\underline{Y}'\underline{L}) = \underline{L}$

where

$$\frac{\partial}{\partial \underline{Y}} = \begin{bmatrix} \frac{\partial}{\partial y_1} \\ \frac{\partial}{\partial y_2} \\ \vdots \\ \frac{\partial}{\partial y_n} \end{bmatrix}$$

If we differentiate the last expression of (A.2) and write $\hat{\underline{\beta}}$ for $\underline{\beta}$, the equations (A.3) are given by

$$-2X'\underline{Y} + 2[X'X]\hat{\underline{\beta}} = 0 \quad \text{i.e.} \quad [X'X]\hat{\underline{\beta}} = X'\underline{Y} \qquad (A.4)$$

Equations (A.4) are referred to as normal equations. When we solve the equations, the estimates are

$$\hat{\underline{\beta}} = [X'X]^{-1}X'\underline{Y} \qquad (A.5)$$

It can be seen that the least-squares estimates $\hat{\beta}_j$ $(j = 1, 2, \ldots, p)$ are linear functions of y_i $(i = 1, 2, \ldots, N)$. Such estimates, which are of the form $A\underline{Y}$, where A is a $p \times n$ matrix, are called linear estimates.

A.4 LEAST-SQUARES ESTIMATES ARE UNBIASED

$$E(\hat{\underline{\beta}}) = E([X'X]^{-1}X'\underline{Y})$$

$$= E\{[X'X]^{-1}X'(X\underline{\beta} + \underline{\epsilon})\}$$

$$= [X'X]^{-1}X'E(X\underline{\beta} + \underline{\epsilon})$$

$$= [X'X]^{-1}X'X\underline{\beta} = \underline{\beta} \qquad\qquad (A.6)$$

A.5 VARIANCE-COVARIANCE MATRIX FOR $\hat{\underline{\beta}}$

$$\text{cov}(\hat{\underline{\beta}}) = E([X'X]^{-1}X'\underline{Y} - \underline{\beta})([X'X]^{-1}X'\underline{Y} - \underline{\beta})'$$

$$= E\{[X'X]^{-1}X'(X\underline{\beta} + \underline{\epsilon}) - \underline{\beta}\}\{[X'X]^{-1}X'(X\underline{\beta} + \underline{\epsilon}) - \underline{\beta}\}'$$

$$= E([X'X]^{-1}X'\underline{\epsilon\epsilon}'X[X'X]^{-1})$$

$$= [X'X]^{-1}X'E(\underline{\epsilon\epsilon}')X[X'X]^{-1}$$

$$= \sigma^2[X'X]^{-1} = \sigma^2 C \qquad\qquad (A.7)$$

where $C = [X'X]^{-1}$. The diagonal elements of the matrix $C = (c_{kj})$, whose (k, j)th element is c_{kj}, denote the <u>variance factors</u>. The variance of $\hat{\beta}_j$ is $\sigma^2 c_{jj}$ $(j = 1, 2, \ldots, p)$. See Sec. 1.3.

A.6 AN UNBIASED ESTIMATE OF σ^2

An unbiased estimate, $\hat{\sigma}^2$ of σ^2 is given by

$$\hat{\sigma}^2 = \frac{1}{N - p}(\underline{Y} - X\hat{\underline{\beta}})'(\underline{Y} - X\hat{\underline{\beta}})$$

$$= \frac{1}{N - p}(\underline{Y} - X[X'X]^{-1}X'\underline{Y})'(\underline{Y} - X[X'X]^{-1}X'\underline{Y})$$

$$= \frac{1}{N - p}\underline{Y}'(I_N - X[X'X]^{-1}X')'(I_N - X[X'X]^{-1}X')\underline{Y}$$

Matrix $A = X[X'X]^{-1}X'$ is _idempotent_, that is, equal to its square, and so is $(I - A)$. Hence, the above reduces to

$$\frac{1}{N-p} \underline{Y}'(I_N - X[X'X]^{-1}X')\underline{Y} = \frac{1}{N-p} (\underline{Y}'\underline{Y} - \hat{\underline{\beta}}'X'\underline{Y}) \qquad (A.8)$$

The second term in expression $(A.8)$ is often convenient for finding the error s.s. Expression $(A.8)$ reduces to

$$\frac{1}{N-p} (X\underline{\beta} + \underline{\epsilon})' (I_N - X[X'X]^{-1}X') (X\underline{\beta} + \underline{\epsilon})$$

$$= \frac{1}{N-p} \underline{\epsilon}'(I_N - X[X'X]^{-1}X')\underline{\epsilon}$$

To show that $\hat{\sigma}^2$ is an unbiased estimate of σ^2, we need the help of the following lemma, which is given below without proof.

Lemma[*] If $\underline{\epsilon}$ is distributed with $E(\underline{\epsilon}) = \underline{0}$ and $E(\underline{\epsilon}\underline{\epsilon}') = \sigma^2 I_N$, then $E(\underline{\epsilon}'A\underline{\epsilon}) = \sigma^2 \text{ Tr } A$.

The term $\hat{\sigma}^2$ is an unbiased estimate of σ^2, since

$$\text{Tr}(I_N - X[X'X]^{-1}X')$$

$$= N - \text{Tr}(X[X'X]^{-1}X')$$

$$= N - \text{Tr}([X'X][X'X]^{-1})$$

$$= N - p$$

When p estimates are provided from N observations, $N - p$ degrees of freedom are left for estimating the variance (see Appendix B).

---

[*] Expected value of a quadratic form is available in a more generalized form.

A.7 GAUSS-MARKOFF THEOREM

The following is the Gauss-Markoff theorem.

Theorem In the general linear hypothesis model of full rank, $\underline{Y} = X\underline{\beta} + \underline{\epsilon}$, where $E(\underline{\epsilon}) = \underline{0}$ and $E(\underline{\epsilon}\underline{\epsilon}') = \sigma^2 I_N$, the best (in the sense of the minimum variance) linear unbiased estimate of $\underline{\beta}$ is provided by least squares, that is, by $\hat{\underline{\beta}} = [X'X]^{-1}X'\underline{Y}$.

The import of the theorem is that, given the design matrix X, of all conceivable linear estimates of $\underline{\beta}$, the least-squares estimator has the minimum variance.

We shall now make an important observation. The choice of X lies in our hands. From (A.7) it is evident that the variance factors c_{jj} for the estimates $\hat{\beta}_j$ $(j = 1, 2, \ldots, p)$ depend on X. The choice of X will therefore be the most efficient if each c_{jj} can be made the smallest possible.

A.8 GENERALIZED LEAST SQUARES

As a generalization of the distribution of $\underline{\epsilon}$, let $E(\underline{\epsilon}) = \underline{0}$ and $E(\underline{\epsilon}\underline{\epsilon}') = \Sigma$, where Σ is the variance-covariance matrix of the error vector $\underline{\epsilon}$. This generalized situation can be reduced, by a suitable transformation, to the Gauss-Markoff setup referred to in the previous section.

Let Σ be positive definite.[*] Then there exists a non-singular matrix P such that $P'\Sigma P = I$ and therefore $\Sigma = [PP']^{-1}$. By premultiplying both sides of $\underline{Y} = X\underline{\beta} + \underline{\epsilon}$ by P',

---

[*]Matrix A of the quadratic form $\underline{Y}'A\underline{Y}$ is said to be positive definite if and only if $\underline{Y}'A\underline{Y} > 0$ for all vectors $\underline{Y}$, where $\underline{Y} \neq 0$.

we can transform the model to

$$\underline{Z} = U\underline{\beta} + \underline{\epsilon}^*$$ (A.9)

where $\underline{Z} = P'\underline{Y}$, $U = P'X$, $\underline{\epsilon}^* = P'\underline{\epsilon}$, $E(\underline{\epsilon}^*) = 0$, and $E(\underline{\epsilon}^* \underline{\epsilon}^{*\prime}) =$
$E(P'\underline{\epsilon\epsilon}'P) = P'\Sigma P = I$. Model (A.9) is now reduced to the form
of (A.2). The generalized least-squares estimates are now given
by

$$\begin{aligned}
\hat{\underline{\beta}}_g &= [U'U]^{-1}U'\underline{Z} \\
&= [X'PP'X]^{-1}X'PP'\underline{Y} \\
&= [X'\Sigma^{-1}X]^{-1}X'\Sigma^{-1}\underline{Y}
\end{aligned}$$ (A.10)

It is easily shown that $E(\hat{\underline{\beta}}_g) = \underline{\beta}$ and that $cov(\hat{\underline{\beta}}_g) =$
$[X'\Sigma^{-1}X]^{-1}$. Also, $cov(\hat{\underline{\beta}}_L) = [X'X]^{-1}[X'\Sigma X][X'X]^{-1}$, where $\hat{\underline{\beta}}_L$ denotes the least-squares estimator given by $\hat{\underline{\beta}} = [X'X]^{-1}X'Y$.
$Cov(\hat{\underline{\beta}}_L)$ is different from $cov(\hat{\underline{\beta}})$ referred to earlier,
since, under the present setup, $E(\underline{\epsilon\epsilon}')$ is different, being equal
to Σ.

Incidentally, when the design matrix X is square and non-
singular, $\hat{\underline{\beta}}_L = \hat{\underline{\beta}}_g$ and $cov(\hat{\underline{\beta}}_L) = cov(\hat{\underline{\beta}}_g)$ (see Sec. 4.12 and
Banerjee [19]).

A.9 THE SITUATION WHEN THE DESIGN MATRIX X
IS NOT OF FULL RANK

When the design matrix X is not of full rank, $[X'X]$ is
singular. Therefore, $[X'X]^{-1}$ will not exist. The normal equa-
tions $[X'X]\hat{\underline{\beta}} = X'\underline{Y}$ will not provide unique solutions, and as
such it will not be possible to have a unique unbiased estimate
for each of the parameter components β_j $(j = 1, 2, \ldots, p)$.

In such a situation, it is possible, however, to have a unique unbiased estimator for a linear function of the parameters β_j ($j = 1, 2, \ldots, p$) under the conditions stated below:

Theorem Let $\underline{\lambda}$ be a vector of known constants. Then, the linear combination $\underline{\lambda}'\underline{\beta}$ will have a unique unbiased estimate if and only if there exists a solution for $\underline{r}$ in the equations $[X'X]\underline{r} = \underline{\lambda}$. The estimate, given by $\underline{r}'X'\underline{Y}$, where $\underline{r}$ is any solution, remains the same and will have the minimum variance.

The choice of $\underline{\lambda}$ is specified in the above theorem. In order that the estimate $\underline{r}'X'\underline{Y}$ of $\underline{\lambda}'\underline{\beta}$ exist, the equations $[X'X]\underline{r} = \underline{\lambda}$ must be consistent, that is, the equations must have a solution $\underline{r}$.

When X is of full rank, the best (in the sense of minimum variance) unbiased estimate of $\underline{\lambda}'\underline{\beta}$ is given by $\underline{\lambda}'\hat{\underline{\beta}} = \underline{\lambda}'[X'X]^{-1}X'\underline{Y}$. The variance of this estimate is given by

$$E(\underline{\lambda}'[X'X]^{-1}X'\underline{Y} - \underline{\lambda}'\underline{\beta})(\underline{\lambda}'[X'X]^{-1}X'\underline{Y} - \underline{\lambda}'\underline{\beta})'$$

$$= E\{\underline{\lambda}'[X'X]^{-1}X'(X\underline{\beta} + \underline{\epsilon}) - \underline{\lambda}'\underline{\beta}\}\{\underline{\lambda}'[X'X]^{-1}X'(X\underline{\beta} + \underline{\epsilon}) - \underline{\lambda}'\underline{\beta}\}'$$

$$= E(\underline{\lambda}'[X'X]^{-1}X'\underline{\epsilon}\epsilon'X[X'X]^{-1}\underline{\lambda})$$

$$= \sigma^2\underline{\lambda}'[X'X]^{-1}\underline{\lambda} \quad \text{since} \quad E(\underline{\epsilon}\epsilon') = \sigma^2 I_N$$

$$= \sigma^2\underline{\lambda}'S^{-1}\underline{\lambda} \qquad\qquad\qquad (A.11)$$

where $S = [X'X]$.

When, however, X is not of full rank, the variance of $\underline{r}'X'\underline{Y}$ is equal to $\sigma^2\underline{\lambda}'S^{-}\underline{\lambda}$, where S^{-} is a g-inverse of $S = [X'X]$ (see Sec. 4.6 and Banerjee [12, 15], Hazra [46], Hazra and Banerjee [47], and Raghavarao [62]).

For a comprehensive study of such models, the reader should consult Graybill [42] and Rao [65].

Appendix B

WEIGHING DESIGNS ILLUSTRATED

This appendix illustrates the working principles of weighing designs with reference to simple numerical examples.   It
is meant primarily for research workers in different disciplines
who might like to use weighing designs in routine work requiring the determination of weights of light objects.

B.1 CHEMICAL BALANCE: THE ARITHMETIC REQUIRED TO
PROVIDE THE ESTIMATES OF WEIGHTS

Let us suppose that there are four light objects denoted
by β_1, β_2, β_3, and β_4, the weights of which are to be determined. Given that only four weighing operations can be made
for the purpose, the best chemical balance (two-pan balance)
design is given by

$$X = \begin{bmatrix} +1 & +1 & +1 & +1 \\ +1 & -1 & +1 & -1 \\ +1 & +1 & -1 & -1 \\ +1 & -1 & -1 & +1 \end{bmatrix}$$

109

and the design model $\underline{Y} = X\underline{\beta} + \underline{\epsilon}$, given in full, looks like

$$
\begin{bmatrix} y_1 \\ y_2 \\ y_3 \\ y_4 \end{bmatrix} = \begin{bmatrix} +1 & +1 & +1 & +1 \\ +1 & -1 & +1 & -1 \\ +1 & +1 & -1 & -1 \\ +1 & -1 & -1 & +1 \end{bmatrix} \begin{bmatrix} \beta_1 \\ \beta_2 \\ \beta_3 \\ \beta_4 \end{bmatrix} + \begin{bmatrix} \epsilon_1 \\ \epsilon_2 \\ \epsilon_3 \\ \epsilon_4 \end{bmatrix}
$$

where y_1, y_2, y_3, and y_4 are the elements of the vector $\underline{Y}$ (the records of weighings), β_1, β_2, β_3, and β_4 are the elements of the vector $\underline{\beta}$ (the four objects, the weights of which are to be determined), and ϵ_1, ϵ_2, ϵ_3, and ϵ_4 are the elements of the experimental error vector $\underline{\epsilon}$. Since the first row of the design matrix X consists entirely of +1 elements, all four objects have to be placed on the left pan according to convention. The balancing weight is naturally placed on the right pan. This is denoted by y_1. Because the second row of X has the elements +1, -1, +1, -1, the first and the third objects, β_1 and β_3, are placed on the left pan, while the second and the fourth objects, β_2 and β_4 are placed on the right pan. The balancing weight, denoted by y_2, is placed on the appropriate pan. Similarly, two more weighing operations are made corresponding to the third and the fourth rows.

If the balancing weight is placed on the right pan, the sign of the record of weighing y is taken as positive. If, on the contrary, the balancing weight is placed on the left pan, the sign of the corresponding record of weighing is negative. Thus, the sign of y_1 in this case is positive because the balancing weight has to be placed on the right pan. The signs of the other records of weighings depend on which pan the balancing weight is placed on in each case.

It should be mentioned that, in this text, the word "weight" is used to denote the record of weighing as well as the weight

piece that comes with the balance. It will be clear from the
context in which sense the word "weight" is used. Written explicitly, the above equations look like

$$y_1 = \beta_1 + \beta_2 + \beta_3 + \beta_4 + \epsilon_1$$
$$y_2 = \beta_1 - \beta_2 + \beta_3 - \beta_4 + \epsilon_2$$
$$y_3 = \beta_1 + \beta_2 - \beta_3 - \beta_4 + \epsilon_3$$
$$y_4 = \beta_1 - \beta_2 - \beta_3 + \beta_4 + \epsilon_4$$

The normal equations (or the estimating equations) for providing the least-squares estimates of the weights, in matrix
notation, are

$$[X'X]\hat{\underline{\beta}} = X'\underline{Y}$$

Since, by assumption, X is nonsingular, the estimates $\hat{\underline{\beta}}$ are
given by

$$\hat{\underline{\beta}} = [X'X]^{-1}X'\underline{Y} = X^{-1}(X')^{-1}X'\underline{Y} = X^{-1}\underline{Y}$$

The above shows that, for such a nonsingular square design matrix,
the least-squares estimates are the same linear combinations of
the y's as those that would be obtained by solving the linear
equations $X\underline{\beta} = \underline{Y}$ for $\underline{\beta}$.

Further, this design matrix X is orthogonal (in the sense
that [X'X] is diagonal). In this case, $[X'X] = 4I_4$, where
I_4 is the identity matrix of order 4. A matrix with this property is called a Hadamard matrix. A Hadamard matrix has the
property that its determinant has the maximum possible value,
considered among all determinants of the same order and formed
with elements x_{ij}, such that $-1 \leq x_{ij} \leq 1$.

From the diagonal property of [X'X], we readily see that
$X^{-1} = X'/4$, indicating that the inverse of the matrix is equal
to its transpose divided by 4. In general, $X^{-1} = X'/N$, when
an $N \times N$ Hadamard matrix H_N exists.

For this design matrix H_4, the estimates $\hat{\underline{\beta}}$ are given by

$$\hat{\underline{\beta}} = X'\underline{y}/4 \qquad\qquad\qquad (B.1)$$

where X', the transpose of X, is given by

$$X' = \begin{bmatrix} +1 & +1 & +1 & +1 \\ +1 & -1 & +1 & -1 \\ +1 & +1 & -1 & -1 \\ +1 & -1 & -1 & +1 \end{bmatrix}$$

Incidentally, in this case, the Hadamard matrix X is symmetric, X' being the same as X. In general, a Hadamard matrix does not have this property. The estimates in (B.1) therefore reduce to

$$\begin{bmatrix} \hat{\beta}_1 \\ \hat{\beta}_2 \\ \hat{\beta}_3 \\ \hat{\beta}_4 \end{bmatrix} = \frac{1}{4} \begin{bmatrix} +1 & +1 & +1 & +1 \\ +1 & -1 & +1 & -1 \\ +1 & +1 & -1 & -1 \\ +1 & -1 & -1 & +1 \end{bmatrix} \begin{bmatrix} y_1 \\ y_2 \\ y_3 \\ y_4 \end{bmatrix}$$

Written explicitly, the estimates are given by

$$\begin{aligned} \hat{\beta}_1 &= (y_1 + y_2 + y_3 + y_4)/4 \\ \hat{\beta}_2 &= (y_1 - y_2 + y_3 - y_4)/4 \\ \hat{\beta}_3 &= (y_1 + y_2 - y_3 - y_4)/4 \\ \hat{\beta}_4 &= (y_1 - y_2 - y_3 + y_4)/4 \end{aligned} \qquad (B.2)$$

Notice that the linear functions (B.2) giving the estimates are mutually orthogonal.

For the purpose of providing the estimates of weights as a routine measure when an H_N is used as a chemical balance weighing design, the following rule can now be stated:

<u>Rule I</u>   Multiply the elements of a column vector of X by
the corresponding elements of the vector $\underline{Y}$, sum up the products,
and divide the sum by N. The operations with successive columns provide the estimates of weights of the successive objects.

B.1.1 A Numerical Example

To illustrate the arithmetic, we furnish below, side by
side, the design matrix H_4 and the records of weighings (in
grams) of four weighing operations:

$$
\begin{bmatrix}
+1 & +1 & +1 & +1 \\
+1 & -1 & +1 & -1 \\
+1 & +1 & -1 & -1 \\
+1 & -1 & -1 & +1
\end{bmatrix}
\qquad
\begin{array}{l}
\text{grams} \\
2.10 = y_1 \\
0.28 = y_2 \\
-0.16 = y_3 \\
0.08 = y_4
\end{array}
$$

Notice that y_3 is negative, implying that the balancing
weight in the third weighing operation was placed on the left
pan. Because the elements of the first column of X are +1,
+1, +1, +1, the estimated weight of the first object is given
by

$$\hat{\beta}_1 = \frac{1}{4}\,[(+1)(y_1) + (+1)(y_2) + (+1)(y_3) + (+1)(y_4)]$$

$$= \frac{1}{4}\,(2.10 + 0.28 - 0.16 + 0.08)$$

$$= 0.575 \text{ g}$$

Similarly, multiplying the elements of the second column with
the corresponding y's, summing up the products, and dividing
the sum by 4, we can estimate the weight of the second object
by

$$\hat{\beta}_2 = \frac{1}{4}\,[(+1)(y_1) + (-1)(y_2) + (+1)(y_3) + (-1)(y_4)]$$

$$= \frac{1}{4}\,(2.10 - 0.28 - 0.16 - 0.08)$$

$$= 0.395 \text{ g}$$

If we proceed as above according to Rule I, $\hat{\beta}_3$ and $\hat{\beta}_4$ are obtained as 0.615 and 0.515 g, respectively.

B.1.2 When There Is a Bias in the Chemical Balance

If it is known that the given chemical balance has a bias, the same design matrix $X = H_4$ may be used to determine the weights of three light objects and that of the bias, which is now considered as an additional object (invisible), the weight (or magnitude) of which is to be determined. Thus, the estimation of the weights of virtually four light objects is involved. However, of these four objects, the bias is such that it will appear as one of the objects in all combinations of the objects.

Let us suppose that, in this case, the bias of the balance is such that a small balancing weight has to be placed on the right pan for the zero correction. This implies that the "invisible" object (bias) is always placed on the left pan and will appear in all combinations with other objects. Let β_1 correspond to the bias, and let the three other objects be denoted by β_2, β_3, and β_4 as before.

Since β_1 (the bias) is always placed on the left pan, all the elements of the first column of X will have to be equal to +1. Notice that the first column of H_4, as adopted, satisfies this condition.

In practice, one would weigh only the three objects β_2, β_3, and β_4 in the combinations indicated by the last three elements of the rows and take the records of weighings for such combinations only. By design, β_1 is automatically included in each of such weighings.

The results of recorded weighings are indicated below, a-
long with the last three columns of the design matrix X.

$$
\text{grams}
\begin{bmatrix} +1 & +1 & +1 \\ -1 & +1 & -1 \\ +1 & -1 & -1 \\ -1 & -1 & +1 \end{bmatrix}
\begin{bmatrix} 1.56 \\ -0.26 \\ -0.71 \\ -0.45 \end{bmatrix}
$$

The weight of 1.56 g was placed on the right pan to balance
the weight of the objects β_2, β_3, and β_4 placed on the left
pan. In the second weighing operation, β_3 was placed on the
left pan, while β_2 and β_4 were placed on the right pan. The
weight of 0.26 g was placed on the left pan to balance. In a
similar manner, the other weighing operations were made, and the
weights were recorded as usual.

If we are not interested in estimating the magnitude of the
bias, the above design matrix (with the last three columns) and
the records of weighings are sufficient to determine the weights
of the three objects. If we apply Rule I, the estimated weight
of β_2 is given by

$$\hat{\beta}_2 = \tfrac{1}{4} [(+1)(1.56) + (-1)(-0.26) + (+1)(-0.71) + (-1)(-0.45)]$$
$$= 0.390 \text{ g}$$

In a similar manner, the estimated weights of β_3 and β_4 are
obtained as 0.615 and 0.520 g, respectively.

If we are also interested in the estimate of the bias, it
can be obtained from an application of Rule I as

$$\hat{\beta}_1 = \tfrac{1}{4} [(+1)(1.56) + (+1)(-0.26) + (+1)(-0.71) + (+1)(-0.45)]$$
$$= 0.035 \text{ g}$$

In an estimation of the weights in a biased balance, we proceeded as if the complete design matrix had been used, although, in practice, we used only the last three columns corresponding to the last three objects β_2, β_3, and β_4.

If some of the elements of the first column of a given design matrix of the above type (Hadamard matrix) are +1's, and some are -1's, the -1's can be changed to +1's by changing all the signs of the corresponding rows. Such a change will not alter the orthogonality of the design matrix or its efficiency as a weighing design.

It should be mentioned that one could use an H_4 to determine the weights of p light objects, where $p \leq 4$.

B.1.3 Standard Error of an Estimated Weight

We have used in the preceding sections an H_4 as a chemical balance design. For a larger number of objects a higher dimensional Hadamard matrix may be used, if it exists. A necessary condition for the existence of a Hadamard matrix H_N of order N is that N be equal to $4t$, where t is a positive integer. In general, when a Hadamard matrix H_N is used as a chemical balance weighing design to determine the weights of N light objects, such a design has the maximum possible efficiency in that the variance of each estimated weight is the minimum possible, being equal to σ^2/N. This derives from the fact that $cov(\hat{\underline{\beta}}) = \sigma^2[X'X]^{-1} = (\sigma^2/N)I_N$. It should be pointed out again that an H_N can be used to determine the weights of p light objects $(p \leq N)$ by utilizing p columns of H_N. In such a situation also, the variance of each estimated weight remains the same, namely, σ^2/N.

If c_{ii} is the ith diagonal element of $C = [X'X]^{-1}$, then $c_{ii}\sigma^2$ is the variance of the ith estimated weight (c_{ii} are variance factors). In the numerical illustration of the preceding section, the variance of each estimated weight is $\sigma^2/4$.

At times, it may also be necessary to indicate the estimate
of standard error for each estimated weight. For this purpose,
it would be necessary, in the first place, to provide an estimate of σ^2. But because four weighing operations were made to
determine the weights of four light objects, no degrees of freedom were left, in this case, for estimating the error variance
σ^2. In order to provide an estimate of the error variance, the
number of objects has to be less than the number of weighing
operations. In general, if N weighing operations are made to
determine the weights of p light objects $(p \leq N)$, then $N - p$
degrees of freedom are left for estimating the error variance
σ^2. For example, if the same H_4 were used to find the weights
of two light objects, $4 - 2 = 2$ degrees of freedom would be
available for estimating the error variance. Let us refer to
the same simple, hypothetical illustration to show the arithmetic needed for estimating the error variance. (The procedure
is exactly the same as that adopted in estimating the error
variance within the frame of such a model.)

Let the second and third columns of the same H_4 be utilized to determine the weights of two light objects, namely, β_2
and β_3, and let the records of weighings be as follows:

$$
\begin{bmatrix} +1 & +1 \\ -1 & +1 \\ +1 & -1 \\ -1 & -1 \end{bmatrix}
\qquad
\begin{array}{c} \text{grams} \\ \begin{bmatrix} 1.02 \\ 0.21 \\ -0.20 \\ -1.00 \end{bmatrix} \end{array}
$$

Applying Rule I, we obtain the estimates of the weights as

$$\hat{\beta}_2 = \frac{1}{4}(1.02 - 0.21 - 0.20 + 1.00)$$

$$= 0.403 \text{ g}$$

$$\hat{\beta}_3 = \frac{1}{4}(1.02 + 0.21 + 0.20 + 1.00)$$

$$= 0.608 \text{ g}$$

In these calculations, the estimates of the weights are worked
out to three places of the decimal only to illustrate the arithmetic of calculating the estimated standard error. The errors
and their squares are as follows:

	Errors	Square of errors
1.02 - 1.011 =	0.009	0.000081
0.21 - 0.205 =	0.005	0.000025
-0.20 + 0.205 =	0.005	0.000025
-1.00 + 1.011 =	0.011	0.000121
		0.000252

In this illustration, all the errors are positive. The errors
are usually both positive and negative.

Since in this illustration, $N - p = 2$, the estimate of
the variance is $0.000252/2 = 0.000126$. Its positive square
root is 0.011 g. Since $N = 4$, the estimated standard error
is $0.011/\sqrt{4} = 0.005$.

B.2 SPRING BALANCE WEIGHING DESIGNS: THE ARITHMETIC
REQUIRED FOR THE MOST EFFICIENT SPRING
BALANCE WEIGHING DESIGNS

In a spring balance (one-pan balance), only one pan is
available for placing the objects. Thus, a balance of which
only one pan can be used for placing the objects is a spring
balance for purposes of weighing designs. Because only one
pan can be used, the elements of the design matrix can assume
only the values +1 and 0; +1 indicates that the corresponding
object is weighed in the combination, and 0 indicates that
the corresponding object is not included in the combination.

The best spring balance weighing design to determine the
weights of three light objects with three weighing operations
is given below:

$$X = \begin{bmatrix} +1 & +1 & 0 \\ +1 & 0 & +1 \\ 0 & +1 & +1 \end{bmatrix}$$

The matrix equation of the model $\underline{Y} = X\underline{\beta} + \underline{\epsilon}$ for this design
can be written as

$$\begin{bmatrix} y_1 \\ y_2 \\ y_3 \end{bmatrix} = \begin{bmatrix} +1 & +1 & 0 \\ +1 & 0 & +1 \\ 0 & +1 & +1 \end{bmatrix} \begin{bmatrix} \beta_1 \\ \beta_2 \\ \beta_3 \end{bmatrix} + \begin{bmatrix} \epsilon_1 \\ \epsilon_2 \\ \epsilon_3 \end{bmatrix}$$

The rows of the above design matrix show that β_1 and β_2, β_1
and β_3, and β_2 and β_3 are to be weighed together in three
successive weighing operations. The records of all the weighings, y_1, y_2, and y_3, will have positive signs.

The normal equations (estimating equations) for providing
the least-squares estimates of the weights are

$$[X'X]\underline{\hat{\beta}} = X'\underline{Y}$$

that is,

$$\begin{bmatrix} 2 & 1 & 1 \\ 1 & 2 & 1 \\ 1 & 1 & 2 \end{bmatrix} \begin{bmatrix} \hat{\beta}_1 \\ \hat{\beta}_2 \\ \hat{\beta}_3 \end{bmatrix} = \begin{bmatrix} 1 & 1 & 0 \\ 1 & 0 & 1 \\ 0 & 1 & 1 \end{bmatrix} \underline{Y} = \begin{bmatrix} y_1 + y_2 \\ y_1 + y_3 \\ y_2 + y_3 \end{bmatrix}$$

Written in full, the equations are

$$2\hat{\beta}_1 + \hat{\beta}_2 + \hat{\beta}_3 = y_1 + y_2$$

$$\hat{\beta}_1 + 2\hat{\beta}_2 + \hat{\beta}_3 = y_1 + y_3$$

$$\hat{\beta}_1 + \hat{\beta}_2 + 2\hat{\beta}_3 = y_2 + y_3$$

These equations have a symmetry. The solutions are therefore easily obtainable as

$$\hat{\beta}_1 = (y_1 + y_2 - y_3)/2$$

$$\hat{\beta}_2 = (y_1 - y_2 + y_3)/2$$

$$\hat{\beta}_3 = (-y_1 + y_2 + y_3)/2$$

It can be seen that the estimates are not orthogonal linear functions of the observations, whereas when an H_4 is used as a chemical balance design, the estimates are available as orthogonal linear functions of the observations.

The variances of the estimated weights can be obtained from the diagonal elements of $cov(\hat{\underline{\beta}}) = [X'X]^{-1} \sigma^2$. However, because the estimates are already provided as linear functions of the observations, the variance of each of the estimated weights is easily obtained as $(3/4)\sigma^2$. This is known to be the minimum possible value of the variance that can be attained in a spring balance in estimating the weights of three light objects with three weighing operations.

The above design matrix of order 3 is denoted by L_3 (see Secs. 3.3 and 3.4 and Mood [56]). In general, when an L_N is used as a weighing design, the normal equations take a symmetrical form, and the solutions of $\hat{\underline{\beta}}$ are easily obtained. By a little algebraic manipulation, the estimated weight of the ith object is given by

$$\hat{\beta}_i = \frac{1}{r - \lambda} [Z_i - (t/2)]$$

where Z_i $(i = 1, 2, \ldots, p)$ are the elements of the ith vector of $X'\underline{Y}$, and $t = \Sigma_{i=1}^{N} y_i$.

With further simplification of the algebraic expressions giving the solutions, the solutions take an even simpler form, as stated in Rule II.

<u>Rule II</u>   To determine the weight of the ith object with L_N as the spring balance design, take the elements of the ith column of L_N, write the corresponding y with a plus sign where there is a +1, and the corresponding y with a minus sign where there is a 0 in the column, and take the algebraic sum and divide the sum by $(N + 1)/2$.

To explain the rule, let us refer to L_3, when $N = 3$ For $\hat{\beta}_1$, the algebraic sum is $(y_1 + y_2 - y_3)$, showing that we have written y_1 and y_2 with a plus sign against the first two elements, which are +1's, and have written y_3 with a minus sign against the third element which is 0. The algebraic sum has been divided by $(N + 1)/2$, which is 2 in this case. The rule may similarly be applied to the second and third columns of L_3 to find $\hat{\beta}_2$ and $\hat{\beta}_3$.

B.2.1 A Numerical Example

To determine the weights of seven light objects $(\beta_1, \beta_2, \ldots, \beta_7)$ with seven weighing operations, the best spring balance design is given by an L_7 (see Secs. 3.3 and 3.4 and Mood [56]). The design matrix and the seven records of weighings are shown below:

$$X = L_7 = \begin{bmatrix} +1 & 0 & +1 & 0 & +1 & 0 & +1 \\ 0 & +1 & +1 & 0 & 0 & +1 & +1 \\ 0 & 0 & 0 & +1 & +1 & +1 & +1 \\ +1 & +1 & 0 & 0 & +1 & +1 & 0 \\ 0 & +1 & +1 & +1 & +1 & 0 & 0 \\ +1 & 0 & +1 & +1 & 0 & +1 & 0 \\ +1 & +1 & 0 & +1 & 0 & 0 & +1 \end{bmatrix}$$

grams

$$\underline{Y} = \begin{bmatrix} y_1 = 2.52 \\ y_2 = 2.20 \\ y_3 = 3.45 \\ y_4 = 2.55 \\ y_5 = 2.80 \\ y_6 = 2.70 \\ y_7 = 2.85 \end{bmatrix}$$

Since $N = 7$, the dividing factor is $(N + 1)/2 = 4$. According to Rule II, the estimates are given by

$$\hat{\beta}_1 = (y_1 - y_2 - y_3 + y_4 - y_5 + y_6 + y_7)/4$$

$$= (2.52 - 2.20 - 3.45 + 2.55 - 2.80 + 2.70 + 2.85)/4$$

$$= 0.54 \text{ g}$$

$$\hat{\beta}_2 = (-y_1 + y_2 - y_3 + y_4 + y_5 - y_6 + y_7)/4$$

$$= (-2.52 + 2.20 - 3.45 + 2.55 + 2.80 - 2.70 + 2.85)/4$$

$$= 0.43 \text{ g}$$

and so on.

As can be seen from the linear expressions estimating the weights, the variance of each estimated weight is $(7/16)\sigma^2$.

In a similar manner, a BIBD can also be used as a spring balance weighing design (see Sec. 3.2 and Banerjee [1]).

APPENDIX C

A LIST OF WEIGHING DESIGNS[*]

C.1 SOME OPTIMUM CHEMICAL BALANCE DESIGNS

In these designs, X denotes the design matrix; N, the number of weighing operations; p, the number of light objects, and $\{c_{ii}\}$, the set of variance factors ($i = 1, 2, \ldots, p$). The variance of ith estimated weight is $c_{ii}\sigma^2$. The covariance terms are not shown.

$$N = p = 2 \qquad\qquad N = p = 3$$

$$X = \begin{bmatrix} +1 & +1 \\ +1 & -1 \end{bmatrix} \qquad X = \begin{bmatrix} +1 & +1 & -1 \\ +1 & -1 & +1 \\ -1 & +1 & +1 \end{bmatrix}$$

$$\{c_{ii}\} = \left\{ \frac{1}{2}, \frac{1}{2} \right\} \qquad\qquad \{c_{ii}\} = \left\{ \frac{1}{2}, \frac{1}{2}, \frac{1}{2} \right\}$$

---

[*]This list is provided for illustrative purposes and is not exhaustive.

$$N = p = 4 \qquad\qquad N = p = 5$$

$$(X = H_4) \quad X = \begin{bmatrix} +1 & +1 & +1 & +1 \\ +1 & -1 & +1 & -1 \\ +1 & +1 & -1 & -1 \\ +1 & -1 & -1 & +1 \end{bmatrix} \qquad X = \begin{bmatrix} +1 & +1 & +1 & +1 & -1 \\ +1 & +1 & +1 & -1 & +1 \\ +1 & +1 & -1 & +1 & +1 \\ +1 & -1 & +1 & +1 & +1 \\ -1 & +1 & +1 & +1 & +1 \end{bmatrix}$$

$$\{c_{ii}\} = \left\{ \frac{1}{4}, \frac{1}{4}, \ldots, \frac{1}{4} \right\} \qquad \{c_{ii}\} = \left\{ \frac{2}{9}, \frac{2}{9}, \ldots, \frac{2}{9} \right\}$$

$$N = p = 6 \qquad\qquad N = p = 6$$

$$X = \begin{bmatrix} +1 & -1 & -1 & -1 & -1 & -1 \\ +1 & -1 & -1 & -1 & +1 & +1 \\ +1 & -1 & -1 & +1 & +1 & -1 \\ +1 & -1 & +1 & +1 & -1 & +1 \\ +1 & +1 & +1 & -1 & +1 & -1 \\ +1 & +1 & -1 & +1 & -1 & +1 \end{bmatrix} \qquad X = \begin{bmatrix} 0 & +1 & +1 & +1 & +1 & +1 \\ +1 & 0 & +1 & -1 & +1 & -1 \\ +1 & +1 & 0 & +1 & -1 & -1 \\ +1 & -1 & +1 & 0 & -1 & +1 \\ +1 & +1 & -1 & -1 & 0 & +1 \\ +1 & -1 & -1 & +1 & +1 & 0 \end{bmatrix}$$

$$\{c_{ii}\} = \left\{ \frac{1}{5}, \frac{1}{5}, \ldots, \frac{1}{5} \right\} \qquad \{c_{ii}\} = \left\{ \frac{1}{5}, \frac{1}{5}, \ldots, \frac{1}{5} \right\}$$

$$N = p = 8; \; p \le 8$$

$$(X = H_8) \quad X = \begin{bmatrix} +1 & +1 & +1 & +1 & +1 & +1 & +1 & +1 \\ +1 & -1 & +1 & -1 & +1 & -1 & +1 & -1 \\ +1 & +1 & -1 & -1 & +1 & +1 & -1 & -1 \\ +1 & -1 & -1 & +1 & +1 & -1 & -1 & +1 \\ +1 & +1 & +1 & +1 & -1 & -1 & -1 & -1 \\ +1 & -1 & +1 & -1 & -1 & +1 & -1 & +1 \\ +1 & +1 & -1 & -1 & -1 & -1 & +1 & +1 \\ +1 & -1 & -1 & +1 & -1 & +1 & +1 & -1 \end{bmatrix}$$

$$\{c_{ii}\} = \left\{ \frac{1}{8}, \frac{1}{8}, \ldots, \frac{1}{8} \right\}$$

$$N = p = 10$$

$$X = \begin{bmatrix} 0 & +1 & +1 & +1 & +1 & +1 & +1 & +1 & +1 & +1 \\ +1 & 0 & +1 & -1 & +1 & -1 & +1 & -1 & +1 & -1 \\ +1 & +1 & 0 & +1 & -1 & +1 & +1 & -1 & -1 & -1 \\ +1 & -1 & +1 & 0 & -1 & +1 & -1 & -1 & +1 & +1 \\ +1 & +1 & -1 & -1 & 0 & +1 & -1 & +1 & +1 & -1 \\ +1 & -1 & +1 & +1 & +1 & 0 & -1 & +1 & -1 & -1 \\ +1 & +1 & +1 & -1 & -1 & -1 & 0 & +1 & -1 & +1 \\ +1 & -1 & -1 & -1 & +1 & +1 & +1 & 0 & -1 & +1 \\ +1 & +1 & -1 & +1 & +1 & -1 & -1 & -1 & 0 & +1 \\ +1 & -1 & -1 & +1 & -1 & -1 & +1 & +1 & +1 & 0 \end{bmatrix}$$

$$\{c_{ii}\} = \left\{\frac{1}{9}, \frac{1}{9}, \ldots, \frac{1}{9}\right\}$$

$$N = p = 12; \ p \leq 12$$

$$(X = H_{12}) \quad X = \begin{bmatrix} +1 & +1 & +1 & +1 & -1 & +1 & -1 & -1 & +1 & +1 & -1 & -1 \\ +1 & -1 & +1 & -1 & -1 & -1 & +1 & -1 & -1 & -1 & -1 & -1 \\ +1 & -1 & -1 & +1 & -1 & +1 & -1 & +1 & -1 & -1 & +1 & -1 \\ +1 & +1 & -1 & -1 & +1 & +1 & +1 & +1 & +1 & -1 & -1 & -1 \\ +1 & -1 & +1 & +1 & +1 & +1 & +1 & +1 & -1 & +1 & -1 & +1 \\ +1 & +1 & -1 & +1 & +1 & -1 & +1 & -1 & -1 & +1 & +1 & -1 \\ +1 & -1 & +1 & -1 & +1 & -1 & -1 & +1 & +1 & +1 & +1 & -1 \\ -1 & -1 & -1 & -1 & +1 & +1 & -1 & -1 & -1 & +1 & -1 & -1 \\ +1 & -1 & -1 & -1 & -1 & +1 & +1 & -1 & +1 & +1 & +1 & +1 \\ -1 & -1 & +1 & +1 & +1 & +1 & +1 & -1 & +1 & -1 & +1 & -1 \\ +1 & -1 & -1 & +1 & +1 & -1 & -1 & -1 & +1 & -1 & -1 & +1 \\ -1 & -1 & -1 & +1 & -1 & -1 & +1 & +1 & +1 & +1 & -1 & -1 \end{bmatrix}$$

$$\{c_{ii}\} = \left\{\frac{1}{12}, \frac{1}{12}, \ldots, \frac{1}{12}\right\}$$

$$N = p = 14; \quad p \leq 14$$

$$
X = \begin{bmatrix}
0 & +1 & +1 & +1 & +1 & +1 & +1 & +1 & +1 & +1 & +1 & +1 & +1 & +1 \\
+1 & 0 & +1 & -1 & +1 & +1 & -1 & -1 & -1 & -1 & +1 & +1 & -1 & +1 \\
+1 & +1 & 0 & +1 & -1 & +1 & +1 & -1 & -1 & -1 & -1 & +1 & +1 & -1 \\
+1 & -1 & +1 & 0 & +1 & -1 & +1 & +1 & -1 & -1 & -1 & -1 & +1 & +1 \\
+1 & +1 & -1 & +1 & 0 & +1 & -1 & +1 & +1 & -1 & -1 & -1 & -1 & +1 \\
+1 & +1 & +1 & -1 & +1 & 0 & +1 & -1 & +1 & +1 & -1 & -1 & -1 & -1 \\
+1 & -1 & +1 & +1 & -1 & +1 & 0 & +1 & -1 & +1 & +1 & -1 & -1 & -1 \\
+1 & -1 & -1 & +1 & +1 & -1 & +1 & 0 & +1 & -1 & +1 & +1 & -1 & -1 \\
+1 & -1 & -1 & -1 & +1 & +1 & -1 & +1 & 0 & +1 & -1 & +1 & +1 & -1 \\
+1 & -1 & -1 & -1 & -1 & +1 & +1 & -1 & +1 & 0 & +1 & -1 & +1 & +1 \\
+1 & +1 & -1 & -1 & -1 & -1 & +1 & +1 & -1 & +1 & 0 & +1 & -1 & +1 \\
+1 & +1 & +1 & -1 & -1 & -1 & -1 & +1 & +1 & -1 & +1 & 0 & +1 & -1 \\
+1 & -1 & +1 & +1 & -1 & -1 & -1 & -1 & +1 & +1 & -1 & +1 & 0 & +1 \\
+1 & +1 & -1 & +1 & +1 & -1 & -1 & -1 & -1 & +1 & +1 & -1 & +1 & 0
\end{bmatrix}
$$

$$\{c_{ii}\} = \left\{ \frac{1}{13}, \; \frac{1}{13}, \; \ldots, \; \frac{1}{13} \right\}$$

$$N = p = 16; \quad p \leq 16$$

$$
(X = H_{16}) \; X = \begin{bmatrix}
+1 & +1 & +1 & +1 & +1 & +1 & +1 & +1 & +1 & +1 & +1 & +1 & +1 & +1 & +1 & +1 \\
+1 & -1 & +1 & -1 & +1 & -1 & +1 & -1 & +1 & -1 & +1 & -1 & +1 & -1 & +1 & -1 \\
+1 & +1 & -1 & -1 & +1 & +1 & -1 & -1 & +1 & +1 & -1 & -1 & +1 & +1 & -1 & -1 \\
+1 & -1 & -1 & +1 & +1 & -1 & -1 & +1 & +1 & -1 & -1 & +1 & +1 & -1 & -1 & +1 \\
+1 & +1 & +1 & +1 & -1 & -1 & -1 & -1 & +1 & +1 & +1 & +1 & -1 & -1 & -1 & -1 \\
+1 & -1 & +1 & -1 & -1 & +1 & -1 & +1 & +1 & -1 & +1 & -1 & -1 & +1 & -1 & +1 \\
+1 & +1 & -1 & -1 & -1 & -1 & +1 & +1 & +1 & +1 & -1 & -1 & -1 & -1 & +1 & +1 \\
+1 & -1 & -1 & +1 & -1 & +1 & +1 & -1 & +1 & -1 & -1 & +1 & -1 & +1 & +1 & -1 \\
+1 & +1 & +1 & +1 & +1 & +1 & +1 & +1 & -1 & -1 & -1 & -1 & -1 & -1 & -1 & -1 \\
+1 & -1 & +1 & -1 & +1 & -1 & +1 & -1 & -1 & +1 & -1 & +1 & -1 & +1 & -1 & +1 \\
+1 & +1 & -1 & -1 & +1 & +1 & -1 & -1 & -1 & -1 & +1 & +1 & -1 & -1 & +1 & +1 \\
+1 & -1 & -1 & +1 & +1 & -1 & -1 & +1 & -1 & +1 & +1 & -1 & -1 & +1 & +1 & -1 \\
+1 & +1 & +1 & +1 & -1 & -1 & -1 & -1 & -1 & -1 & -1 & -1 & +1 & +1 & +1 & +1 \\
+1 & -1 & +1 & -1 & -1 & +1 & -1 & +1 & -1 & +1 & -1 & +1 & +1 & -1 & +1 & -1 \\
+1 & +1 & -1 & -1 & -1 & -1 & +1 & +1 & -1 & -1 & +1 & +1 & +1 & +1 & -1 & -1 \\
+1 & -1 & -1 & +1 & -1 & +1 & +1 & -1 & -1 & +1 & +1 & -1 & +1 & -1 & -1 & +1
\end{bmatrix}
$$

$$\{c_{ii}\} = \left\{ \frac{1}{16}, \; \frac{1}{16}, \; \ldots, \; \frac{1}{16} \right\}$$

C.2 SOME OPTIMUM SPRING BALANCE DESIGNS

$$N = p = 3$$

$$(X = L_3) \quad X = \begin{bmatrix} +1 & +1 & 0 \\ +1 & 0 & +1 \\ 0 & +1 & +1 \end{bmatrix}$$

$$\{c_{ii}\} = \left\{ \frac{3}{4}, \frac{3}{4}, \frac{3}{4} \right\}$$

$$N = p = 4$$

$$X = \begin{bmatrix} +1 & 0 & 0 & +1 \\ +1 & +1 & +1 & 0 \\ 0 & 0 & +1 & +1 \\ 0 & +1 & 0 & +1 \end{bmatrix}$$

$$\{c_{ii}\} = \left\{ \frac{7}{9}, \frac{7}{9}, \frac{7}{9}, \frac{4}{9} \right\}$$

$$N = p = 4$$

$$X = \begin{bmatrix} +1 & +1 & +1 & 0 \\ +1 & +1 & 0 & +1 \\ +1 & 0 & +1 & +1 \\ 0 & +1 & +1 & +1 \end{bmatrix}$$

$$\{c_{ii}\} = \left\{ \frac{7}{9}, \frac{7}{9}, \frac{7}{9}, \frac{7}{9} \right\}$$

$$N = p = 5$$

$$X = \begin{bmatrix} 0 & 0 & 0 & +1 & +1 \\ 0 & 0 & +1 & +1 & 0 \\ 0 & +1 & +1 & 0 & +1 \\ +1 & +1 & 0 & +1 & 0 \\ +1 & 0 & +1 & 0 & +1 \end{bmatrix}$$

$$\{c_{ii}\} = \left\{ \frac{19}{25}, \frac{19}{25}, \frac{16}{25}, \frac{11}{25}, \frac{16}{25} \right\}$$

$$N = p = 6$$

$$X = \begin{bmatrix} +1 & +1 & +1 & 0 & 0 & 0 \\ +1 & 0 & 0 & 0 & +1 & +1 \\ +1 & 0 & 0 & +1 & +1 & 0 \\ 0 & 0 & +1 & +1 & 0 & +1 \\ 0 & +1 & +1 & 0 & +1 & 0 \\ 0 & +1 & 0 & +1 & 0 & +1 \end{bmatrix}$$

$$\{c_{ii}\} = \left\{ \frac{17}{27}, \frac{17}{27}, \ldots, \frac{17}{27} \right\}$$

$$N = p = 7; \; p \le 7$$

$$X = \begin{bmatrix} +1 & 0 & +1 & 0 & +1 & 0 & +1 \\ 0 & +1 & +1 & 0 & 0 & +1 & +1 \\ 0 & 0 & 0 & +1 & +1 & +1 & +1 \\ +1 & +1 & 0 & 0 & +1 & +1 & 0 \\ 0 & +1 & +1 & +1 & +1 & 0 & 0 \\ +1 & 0 & +1 & +1 & 0 & +1 & 0 \\ +1 & +1 & 0 & +1 & 0 & 0 & +1 \end{bmatrix}$$

$(X = L_7;$ symm. BIBD:
$v = b = 7, \; r = k = 4,$
$\lambda = 2)$

$$\{c_{ii}\} = \left\{ \frac{7}{16}, \frac{7}{16}, \ldots, \frac{7}{16} \right\}$$

If the design is used for $p < 7$, the variance factors will be different.

$$N = 10; \; p = 6$$

$$X = \begin{bmatrix} +1 & +1 & 0 & 0 & +1 & 0 \\ +1 & +1 & 0 & 0 & 0 & +1 \\ +1 & 0 & +1 & +1 & 0 & 0 \\ +1 & 0 & +1 & 0 & 0 & +1 \\ +1 & 0 & 0 & +1 & +1 & 0 \\ 0 & +1 & +1 & +1 & 0 & 0 \\ 0 & +1 & +1 & 0 & +1 & 0 \\ 0 & +1 & 0 & +1 & 0 & +1 \\ 0 & 0 & +1 & 0 & +1 & +1 \\ 0 & 0 & 0 & +1 & +1 & +1 \end{bmatrix}$$

$(BIBD: \; v = 6, \; k = 3,$
$r = 5, \; b = 10,$
$\lambda = 2)$

$$\{c_{ii}\} = \left\{ \frac{13}{45}, \frac{13}{45}, \ldots, \frac{13}{45} \right\}$$

$$N = 14; \quad p = 8$$

$$X = \begin{bmatrix}
+1 & +1 & +1 & +1 & 0 & 0 & 0 & 0 \\
+1 & +1 & 0 & 0 & 0 & 0 & +1 & +1 \\
+1 & 0 & +1 & 0 & 0 & +1 & 0 & +1 \\
+1 & 0 & 0 & +1 & 0 & +1 & +1 & 0 \\
0 & 0 & 0 & 0 & +1 & +1 & +1 & +1 \\
0 & 0 & +1 & +1 & +1 & +1 & 0 & 0 \\
0 & +1 & 0 & +1 & +1 & 0 & +1 & 0 \\
0 & +1 & +1 & 0 & +1 & 0 & 0 & +1 \\
+1 & +1 & 0 & 0 & +1 & +1 & 0 & 0 \\
+1 & 0 & +1 & 0 & +1 & 0 & +1 & 0 \\
+1 & 0 & 0 & +1 & +1 & 0 & 0 & +1 \\
0 & 0 & +1 & +1 & 0 & 0 & +1 & +1 \\
0 & +1 & 0 & +1 & 0 & +1 & 0 & +1 \\
0 & +1 & +1 & 0 & 0 & +1 & +1 & 0
\end{bmatrix}$$

(BIBD: $v = 8$,
$k = 4$, $r = 7$,
$b = 14$, $\lambda = 3$)

$$\{c_{ii}\} = \left\{ \frac{25}{112}, \frac{25}{112}, \ldots, \frac{25}{112} \right\}$$

1. Banerjee, K. S. Weighing designs and balanced incomplete blocks. Ann. Math. Stat., 19, 394-399 (1948).

2. Banerjee, K. S. A note on weighing designs. Ann. Math. Stat.,20, 300-304 (1949).

3. Banerjee, K. S. On certain aspects of spring balance designs. Sankhyā, 9, 367-376 (1949).

4. Banerjee, K. S. On the variance factors of weighing designs in between two Hadamard matrices. Calcutta Stat. Ass. Bull., 2, 38-42 (1949).

5. Banerjee, K. S. On the construction of Hadamard matrices. Science and Culture, 14, 434-435 (1949).

6. Banerjee, K. S. Weighing designs. Calcutta Stat. Ass. Bull., 3, 64-76 (1950).

7. Banerjee, K. S. Some contributions to Hotelling's weighing designs. Sankhyā, 10, 371-382 (1950).

8. Banerjee, K. S. How balanced incomplete block designs may be made to furnish orthogonal estimates in weighing designs. Biometrika, 37, 50-58 (1950).

9. Banerjee, K. S. Some observations on practical aspects of weighing designs. Biometrika, 38, 248-251 (1951).

10. Banerjee, K. S. Weighing designs and partially balanced incomplete blocks. Calcutta Stat. Ass. Bull., 4, 36-38 (1952).

11. Banerjee, K. S. On Hotelling's weighing designs under auto-correlation of errors. Ann. Math. Stat., 36, 1829-1834 (1965).

12. Banerjee, K. S. Singularity in Hotelling's weighing designs and a generalized inverse. Ann. Math. Stat., 37, 1021-1032 (1966). (A correction note appears in Ann. Math. Stat., 40, 2, 719.)

13. Banerjee, K. S. On non-randomized fractional weighing designs. Ann. Math. Stat., 37, 1836-1841 (1966).

14. Banerjee, K. S. In justification of Wald's criterion of efficiency of experimental designs in relation to the weighing problem. Aust. J. Stat., 14-2, 139-142 (1970).

15. Banerjee, K. S. Singular weighing designs and a reflexive generalized inverse. J. Amer. Stat. Ass., 67, No. 337, 211-212 (1972).

16. Banerjee, K. S. Some observations on repeated spring balance weighing designs. Ann. Inst. Stat. Math. (Tokyo), in press.

17. Banerjee, K. S. On D-optimability of Yates' orginal example in the Yates-Hotelling weighing problem. Commun. Stat., 3 No. 2, 185-190 (1974).

18. Banerjee, K. S. An introduction to the weighing problem, Chem. Scripta, 6, No. 4, 158-162 (1974).

19. Banerjee, K. S. A new model for the weighing problem and a comment on the construction of balances. In press.

20. Baumert, L. S., Golomb, S. W., and Hall, M. Jr. Discovery of a Hadamard matrix of order 92. Bull. Amer. Math. Soc. 68, 237-238 (1962).

21. Beckman, R. J. Randomized spring balance weighing designs. Unpublished Ph.D. dissertation, Kansas State Univ., 1969.

22. Beckman, R. J. An application of multivariate weighing designs, Commun. Stat., 1, No. 6, 561-565 (1972).

23. Bhaskararao, M. Weighing designs when n is odd. Ann. Math. Stat., 37, No. 5, 1371-1381 (1966).

24. Bose, R. C. On the construction of balanced incomplete block designs. Ann. Eugenics, 9, No. 3, 353-399 (1939).

25. Bose, R. C., and Nair, K. R. Partially balanced incomplete block designs. Sankhya, 4, 337-372 (1939).

26. Box, G. E. P., and Draper, N. R. A basis for selection of a response surface design. J. Amer. Stat. Ass., 54, 622-654 (1959).

27. Box, G. E. P., and Hunter, J. S. Multifactor experimental designs for exploring response surfaces. Ann. Math. Stat., 28, 195-241 (1957).

28. Box, G. E. P., and Wilson, K. B. On the experimental attainment of optimum conditions. *J. Roy. Stat. Soc. Sec B*, 13, 1-45 (1951).

29. Broyles, A. R., and Eckert, R. E. Maximization of the conversion to m-toluenesulfonic acid in sulfuric acid sulfonation. *Ind. Chem. Process Des. Develop.*, 12, No. 3, 296-300 (1973).

30. Chakrabarti, M. C. *Mathematics of Design and Analysis of Experiments*, Chap. 6. New York: Asia Publishing House, 1962.

31. Dey, A. A note on weighing designs. *Ann. Inst. Stat. Math.*, 21, 343-346 (1969).

32. Dey, A. On some chemical balance weighing designs, *Aust. J. Stat.*, 13, No. 3, 137-141 (1971).

33. Dey, A. Some remarks on chemical balance weighing designs. *J. Ind. Soc. Agr. Stat.*, 24, 119-126 (1962).

34. Ehlich, H. Determinantenabeschatzungen für binäre matrizen. *Math. Z.*, 83, 123-132, MR 28 No. 4003 (1964).

35. Ehlich, H. Neue Hadamard Matrizen, *Arch. Math.*, 16, 34-36 (1965).

36. Ehlich, H., and Zeller, K. Binäare Matrizen, *ZAMM*, 42, 20-21 (1962).

37. Ehrenfeld, S. On the efficiencies of experimental designs. *Ann. Math. Stat.*, 26, 247-255 (1955).

38. Federer, W. T. *Experimental Design*, Chap. 15, p. 432. New York: Macmillan, 1955.

39. Finney, D. J. The fractional replication of factorial arrangements. *Ann. Eugenics*, 12, 291-301 (1945).

40. Fisher, R. A., and Yates, F. *Statistical Tables for Biological, Agricultural and Medical Research, 6th ed.* New York: Hafner, 1963.

41. Gordon, P. A maximal determinant. *Amer. Math. Mon.*, 53, 220-221 (1940).

42. Graybill, F. A. *An Introduction to Linear Statistical Models*, Vol. I. New York: McGraw-Hill, 1961.

43. Hadamard, J. Resolution d'une question relative aux determinants. *Bull. des Sci. Math.*, 17, 240-246 (1893).

44. Hall, Marshall, Jr., *Combinatorial Theory*. Waltham, Mass.: Blaisdell, 1967.

45. Harwit, M. Spectrometric imager. *Appl. Opt.*, 10, No. 6, 1415-1421 (1971).

46. Hazra, P. K. On the structure of singular weighing designs, Commun. Stat., 2, No. 6, 575-579 (1973).

47. Hazra, P. K., and Banerjee, K. S. On the augmentation procedure in singular weighing designs. J. Amer. Stat. Ass. 68, 392-393 (1973).

48. Hoerl, A.E., and Kennard, R.W. Ridge regression. Biased estimation for non-orthogonal problems. Technometrics, 12, 55-67 (1970).

49. Hotelling, H. Some improvements in weighing and other experimental techniques. Ann. Math. Stat., 15, 297-306 (1944).

50. Kempthorne, O. The factorial approach to the weighing problem. Ann. Math. Stat., 19, 238-248 (1949).

51. Kiefer, J. On the nonrandomized optimality and randomized nonoptimality of symmetrical designs. Ann. Math. Stat., 29, 675-699 (1958).

52. Kiefer, J. Optimum experimental designs. J. Roy. Stat. Soc. Ser. B, 21, 272-304(1959).

53. Kishen, K. On the design of experiments for weighing and making other types of measurements. Ann. Math. Stat., 16, 294-300 (1945).

54. Kulshreshtha, A. C., and Dey, A. A new weighing design. Aust. J. Stat., 12, 166-168 (1970).

55. Lese, William G., Jr., and Banerjee, K.S. Orthogonal estimates in weighing designs. Proc. 18th Conf. on Designs in Army Research and Testing, 1972. (Originally, Lese's Ph.D. dissertation, Univ. of Delaware, Newark.)

56. Mood, A. M. On Hotelling's weighing problem. Ann. Math. Stat., 17, 432-446 (1946).

57. Moriguti, S. Optimality of orthogonal designs. Report of Statistical Applications Research, Union of Japanese Scientists and Engineers (Tokyo) 3, 1-24 (1954).

58. Paley, R. E. A. C. On orthogonal matrices. J. Math. Phys. (Cambridge Mass.) 12, 311-320 (1933).

59. Plackett, R. L., and Burman, J. P. The design of optimum multifactorial experiments. Biometrika, 33, 305-325 (1946).

60. Raghavarao, D. Some optimum weighing designs. Ann. Math. Stat., 30, 295-303 (1959).

61. Raghavarao, D. Some aspects of weighing designs. Ann. Math. Stat., 31, 878-884 (1960).

62. Raghavarao, D. Singular weighing designs. Ann. Math. Stat., 35, 673-680 (1964).

63. Raghavarao, D. Constructions and Combinatorial Problems
     in Design of Experiments, New York: Wiley, 1971.

64. Rao, C. R. On the most efficient designs in weighing.
     Sankhya, 7, 440 (1946).

65. Rao, C. R. Linear Statistical Inference and Its Applica-
     tions. New York: Wiley, 1965.

66. Sane, P. P., Woods, J. M., and Eckert, R. E. Squential
     design of initial experiments for chemical kinetic model-
     ing--isomerization of n-pentane. Chem. Eng. Sci. 28, 1609-
     1616 (1973).

67. Sihota, S. S., and Banerjee, K. S. Biased estimation in
     weighing designs, Sankhya, Ser. B, 36, Pt. 1, 55-64 (1974).

68. Sloane, N. J. A., Fine, T., Phillips, P. G., and Harwit, M.
     Codes for multiplex spectrometry. Appl. Opt., 8, 2103-2106
     (1969).

69. Sloane, N. J. A., Fine, T., and Phillips, P. G. How the
     use of masks to multiplex the light at the entrance and
     exit slits can enhance the precision of grating spectrometer
     measurements. Opt. Spectra, 4, 50-53 (1970).

70. Wald, A. On the efficient design of statistical investi-
     gations. Ann. Math. Stat., 14, 134-140 (1943).

71. Williamson, J. Hadamard's determinant theorem and the sum
     of four squares. Duke Math. J., 11, 65-82 (1944).

72. Williamson, J. Note on maximal determinants. Amer. Math.
     Mon., 53, 222-224 (1946).

73. Williamson, J. Determinants whose elements are 0 and 1.
     Amer. Math. Mon., 53, 427-434 (1946).

74. Yang, C. H. A construction for maximal (+1,-1)-matrices
     of order n ≡ 2 (mod 4). Math. Mon., 72, 293, MR 32 No.
     5678 (1966).

75. Yang, C. H. Some designs for maximal (+1,-1)-determinants
     of order n ≡ 2 (mod 4). Math. Comp. 20, 147-418 (1966).

76. Yang, C. H. A construction for maximal (+1, -1)-matrix of
     order 54. Bull. Amer. Math. Soc., V, 72, 293 MR 32
     #5678 (1966).

77. Yang, C. H. On designs of maximal (+1, -1)-matrices of
     order n ≡ 2 (mod 4). Math. Comp. 22, 174-180 (1968).

78. Yates, F. Complex experiments. J. Roy. Stat. Soc., Suppl.,
     2, 181-247 (1935).

79. Youden, W. J. Linked blocks: A new class of incomplete
     block designs (Abstract). Biometrics, 7, 124 (1951).

80. Youden, W. J. Evaluation of chemical analyses in two rocks.
     Technometrics, 1, 409-417 (1959).

81. Youden, W. J. Systematic errors in physical constants.
     Technometrics, 4, 111-123 (1962).

82. Youden, W. J., Connor, W. S., and Severo, N. C. Measure-
     ments made by matching with known standards. Technometrics,
     1, 101-109 (1959).

83. Zacks, S. Randomized fractional weighing designs. Ann.
     Math. Stat., 37, 1382-1395 (1966).